AF368948

# Selected Papers from the 6th Fábos Conference on Landscape and Greenway Planning

# Selected Papers from the 6th Fábos Conference on Landscape and Greenway Planning

## Adapting to Expanding and Contracting Cities

Editor

**Richard C. Smardon**

MDPI • Basel • Beijing • Wuhan • Barcelona • Belgrade • Manchester • Tokyo • Cluj • Tianjin

*Editor*
Richard C. Smardon
State University of New York
USA

*Editorial Office*
MDPI
St. Alban-Anlage 66
4052 Basel, Switzerland

This is a reprint of articles from the Special Issue published online in the open access journal *Land* (ISSN 2073-445X) (available at: https://www.mdpi.com/journal/land/special_issues/faboa_conference).

For citation purposes, cite each article independently as indicated on the article page online and as indicated below:

LastName, A.A.; LastName, B.B.; LastName, C.C. Article Title. *Journal Name* **Year**, *Volume Number*, Page Range.

ISBN 978-3-0365-0308-0 (Hbk)
ISBN 978-3-0365-0309-7 (PDF)

# Contents

# About the Editor

**Richard C. Smardon** has worked in academic, government and private practice positions before coming to the SUNY College of Environmental Science and Forestry. He is a SUNY Distinguished Service Professor Emeritus at SUNY/ESF. He has a Ph.D. in Environmental Planning from the University of California, Berkeley, and a Masters in Landscape Architecture and a Bachelors in Environmental Design from the University of Massachusetts, Amherst. He has edited/written seven books, the most recent publications being The Renewable Energy Landscape (2017), Revitalizing Urban Waterway Communities: Streams of Environmental Justice (2018) and Education for Sustainable Human and Environmental Systems (2019). He is on the editorial boards of several journals including Land, Water and Urban Planning plus he is a book review editor for the Landscape Journal and the Journal of Environmental Studies and Sciences.

 *land*

*Editorial*

# 6th Fábos Conference on Landscape and Greenway Planning

**Richard Smardon**

Department of Environmental Studies, SUNY College of Environmental Science and Forestry, Syracuse, NY 13210, USA; rsmardon@esf.edu

Received: 3 November 2020; Accepted: 4 November 2020; Published: 8 November 2020

**Abstract:** This editorial is an overview of a Special Issue of *Land* entitled *"Selected Papers from the 6th Fábos Conference on Landscape and Greenway Planning: Adapting and Expanding Contracting Cities."* This Special Issue of land contains six papers—most of which were presented at the 6th Fábos Conference on Landscape and Greenway Planning (Fábos et al. 2019) held at the University of Massachusetts Amherst 28–30 March 2019.The Fábos conference theme was to explore the social and economic potential of linear green spaces in urban areas that are declining or expanding.

**Keywords:** greenway; planning; urban landscape

## 1. Introduction

Being a former graduate student (MLA 1973) mentored by Emeritus Professor Julius Gy Fábos, I am very pleased to guest edit this Special Issue of *land* entitled *"Selected Papers from the 6th Fábos Conference on Landscape and Greenway Planning: Adapting and Expanding Contracting Cities."* This Special Issue of *land* contains six papers—most of which were presented at the 6th Fábos Conference on Landscape and Greenway Planning [1] held at the University of Massachusetts Amherst 28–30 March 2019.

The Fábos Conference on Landscape and Greenway Planning is held every three years to bring together scholars who are influencing landscape planning, policy making and greenway development from local to international levels. The aim is to explore how landscape architects and planners from different countries have approached greenway planning and to understand how greenways can be tailored to each country or region's geographic, cultural and political contexts. The 2019 Fábos conference theme was to explore the social and economic potential of linear green spaces in urban areas that are declining or expanding.

## 2. Results and Discussion

There are five research papers and one review paper that address this theme. Olivia Horte and Theodore Eisenman' s review paper [2] is a systematic literature review and typology of urban greenway scholarly papers. They reviewed some 52 referred articles to identify gaps in greenway scholarship as well as develop a typology for urban greenway research. The remaining papers address physical, cultural and historical greenway research occurring in Zhengzhou City China, Lisbon Portugal, Amman Jordan, Hungary and Guangzhou China.

Huawei Li et al. [3] map and analyze the role of parks effect on the urban heat island in Zhengzhou City China, which is an expanding urban area. The authors use remote sensing imagery to analyze 123 parks looking at the cooling effect on the dense expanding urban area of Zhengzhou City. In another paper in the Special Issue, Wenxiu Chi and Guangsi Lin conducted a case study [4] on a highly dense area in Guangzhou China to see if the greenway areas match the needs of resident's physical and social activities.

There are two papers in the Special Issue that analyze aspects of urban morphology, which is important for urban neighborhood development and structure. Rui Justo and Maria Matis Silva [5] analyze the role of vegetation in decoding urban morphology within three neighborhoods in Lisbon Portugal. Anne Gharaibeh et al. [6] also look at urban morphology of Amman Jordan with specific reference to the role of urban streams. They utilized historic areal photos and maps as well as interviews to better understand the perception of urban morphology change over time.

Finally Albert Fekete and Lazio Kollányi [7] utilize research based design techniques applied to historic castle gardens in the Caspian Basin Hungary. Such knowledge will be useful for historic interpretation as well as garden restoration.

## 3. Conclusions

These six Special Issue papers explore the physical, historic and cultural aspects of urban greenspace from multidisciplinary perspectives. More conference papers can be found in the online conference proceedings [1].

**Funding:** This research received no external funding.

**Conflicts of Interest:** The author declares no conflict of interest.

## References

1. Fábos, J.G.; Ahern, J.; Breger, B.; Eisenman, T.S.; Jombach, S.; Kollányi, L.; Lindhult, M.S.; Ryan, R.L.; Valánszki, I. Adapting to Expanding and Contracting Cities, Book of Abstracts, 6th Fábos Conference on Landscape and Greenway Planning, March 28–30, 2019, Amherst, MA. In Proceedings of the Fábos Conference on Landscape and Greenway Planning, Amherst, MA, USA, 28–30 March 2019; Volume 6. [CrossRef]
2. Horte, O.S.; Eisenman, T.S. Urban Greenways: A Systematic Review and Typology. *Land* **2020**, *9*, 40. [CrossRef]
3. Li, H.; Wang, G.; Tian, G.; Jombach, S. Mapping and Analyzing the Park Cooling Effect on Urban Heat Island In an Expanding City: A Case Study in Zhengzhou City China. *Land* **2020**, *9*, 57. [CrossRef]
4. Chi, W.; Lin, G. The Use of Community Greenways: A Case Study on a Linear Greenway Space in High Dense Residential Areas, Guangzhou. *Land* **2020**, *8*, 188. [CrossRef]
5. Justo, R.; Silva, M.M. The Role of Vegetation in the Morphological Decoding of Lisbon (Portugal). *Land* **2020**, *9*, 18. [CrossRef]
6. Gharaibeh, A.A.; AlZu'bi, E.M.; Abuhasson, L.B. Amman (City of Waters); Policy, Land Use, and Character Changes. *Land* **2019**, *8*, 195. [CrossRef]
7. Fekete, A.; Kollányi, L. Research-Based Design Approaches in Historic Garden Renovation. *Land* **2019**, *8*, 192. [CrossRef]

**Publisher's Note:** MDPI stays neutral with regard to jurisdictional claims in published maps and institutional affiliations.

 *land*

*Review*

# Urban Greenways: A Systematic Review and Typology

**Olivia S. Horte [1] and Theodore S. Eisenman [2,*]**

[1] Rose Kennedy Greenway Conservancy, Boston, MA 02111, USA; ohorte@rosekennedygreenway.org
[2] Department of Landscape Architecture and Regional Planning, University of Massachusetts-Amherst, Amherst, MA 01003, USA
* Correspondence: teisenman@umass.edu

Received: 31 October 2019; Accepted: 12 December 2019; Published: 1 February 2020

**Abstract:** Greenways are multifunctional linear landscapes that provide a range of socio-ecological benefits. As a domain of landscape planning research, greenways gained traction in the late 20th century and today, there is substantial interest in greenway planning and design. This is especially true in urban areas, as noted at the sixth Fábos Conference on Landscape and Greenway Planning. Yet, cities encompass biophysical flows, sociopolitical relationships, and formal structures that are distinct from non-urban areas and urban greenways may reflect an evolving type of landscape planning and design that is related to but distinct from greenways writ large. To the best of our knowledge, there has been no previous review of scholarship on greenways in an urban context. We address the aforementioned gaps by reporting on a systematic assessment of peer-reviewed literature. The review encompasses 52 refereed articles using the term "urban greenway" or "urban greenways" in the title, abstract, or keywords drawn from three prominent academic databases. Our analysis covers seven research categories, and this undergirds a typology and definition of urban greenways. In so doing, we seek to illuminate typical traits of urban greenways to inform future landscape planning scholarship and practice.

**Keywords:** urban greenways; urban parks; urban greening; green infrastructure; systematic review; landscape typology

---

## 1. Introduction

In scholarly literature, greenways have been defined as "networks of land containing linear elements that are planned, designed and managed for multiple purposes including ecological, recreational, cultural, aesthetic, or other purposes compatible with the concept of sustainable land use" [1]. Early use of the "greenway" term can be traced to Elenor Smith Morris' publication of "New urban design concepts: greenways and movement structure: the Philadelphia plan" in *Architect's Yearbook* [2], and William H. Whyte's *The Last Landscape* [3], which describes greenways as critical linkages and connectors in a hierarchy of urban green spaces. The idea gained further scholarly traction with *Greenways for America* [4] and *Greenways: A Guide to Planning, Design, and Development* [5], followed by two Special Issues of a peer-reviewed journal, *Landscape and Urban Planning*, dedicated to this topic [6–8].

In practice, however, greenway precedents include tree planting along roads and canals, dating back 2000 years in China [9]; landscape corridors dating to ancient Rome; planted boulevards in 18th century European cities; and the 19th century parkways and park systems of U.S. cities [10,11]. And today, there is substantial interest in greenway planning and design. This is especially true in urban areas due to the growing concentration of people in cities [12], and the limited amount of available space in increasingly built up settlements. Greenways may also be of contemporary interest in so-called "legacy" or "shrinking" cities that are characterized by a diminishing population and increasing swaths of vacant land [13,14]. As such, urban greenways may be a form of 21st century landscape planning and design that has the potential to address the challenges and opportunities of both expanding and contracting cities, as noted at the 6th Fábos Conference on Landscape and Greenway Planning at which a preliminary version of this study was presented [15].

In this paper, we treat urban greenways as a related but distinct subset of greenways writ large. We believe this is justified for reasons pertaining to pre-existing condition, location, and extent. Regarding pre-existing condition, greenways are often characterized as undeveloped and environmentally sensitive corridors to be conserved in advance of urbanization [1,5]. This highlights the importance of greenways as a sustainable planning strategy to contain or shape urban expansion, reduce land fragmentation, and maintain "landscape integrity" [16], drawing upon scholarship in landscape ecology [17] and subsequent literature on green infrastructure [18], where greenways are critical links/corridors that connect hubs/patches of natural lands to support biotic and abiotic ecological processes. Stated another way: greenways are critical elements of "nature's...pre-existent...super infrastructure" [19].

Pre-existing condition also relates to location. The Oxford English Dictionary, for example, defines a greenway as "a grassy path or way; a piece of undeveloped land *near* an urban area, set aside for recreational use or environmental conservation" [20]. Regarding extent, greenways are often conceived as a regional, state, national, international or even continental network that can include but generally transcends urban areas [1,11,21–23]. One of the most prominent greenways in the United States, for example, is the Appalachian Trail, which extends some 2190 miles along the eastern mountain chain from which it takes its name; and the experience it is intended to provide is essentially an escape from the urban condition [24].

This is not to suggest that urban areas are not conducive to greenways. Indeed, noteworthy sites in landscape planning history include urban greenways [11]; greenways can constitute a "living network" that provides "people with access to open spaces close to where they live...and link together rural and urban spaces" [25]. Yet, urban areas differ from non-urban areas in important ways, including biophysical flows, sociopolitical dynamics, and formal structure [26–30]. Moreover, in much the same way that urban parks can differ from non-urban parks [31,32], greenways located in highly urbanized areas contend with conditions that can be quite different from non-urban areas [33,34]. Thus, it behooves landscape planners and designers—and associated scholarship—to understand the unique traits of urban greenways to meet the needs of current and future cities.

Towards that goal, this paper addresses the following question: What are the traits that distinguish urban greenway scholarship and practice as an evolving form of landscape planning and design? With this in mind, we address three objectives: (1) illuminate the ways that urban greenways may be a distinct subset of greenways writ large; (2) systematically review scholarly literature on urban greenways; (3) develop a typology and definition of urban greenways. To the best of our knowledge, there has been no previous literature review on greenways in an urban context. Literature reviews are foundational for advancing state-of-the-art understanding of a topic [35,36], and for doing substantive and thorough research [37]. We address this gap and the aforementioned topics by reporting on a systematic literature review of urban greenways and development of an associated typology. Although urban greenways tend to be designed by landscape architects and urban planners, they are also domains of research for a range of fields, including ecology, geography, sociology, wildlife conservation, economics, human health, and others. A systematic review, typology, and definition may, thus, facilitate scholarship on urban greenways across a range of disciplines.

## 2. Methods: A Systematic Review

There are many kinds of literature review and the type of review should be appropriate for the subject and goal at hand [36]. In this case, we conducted a systematic review as there has, to the best of our knowledge been no review on urban greenways, yet greenways are common in urban areas and there is a sizable scholarly literature on this topic. As such, a systematic review is appropriate because it helps to clarify the state of existing research and associated implications for future research [38]. In structuring the methodology of this review, we drew upon systematic review precedents in urban greening and landscape planning [35,39].

We searched for the terms "urban greenway" and "urban greenways" through 2018 within the title, keywords, or abstract of three databases: *Web of Science, ScienceDirect,* and *Avery Index to Architectural Periodicals.* We then eliminated overlapping results across these database searches. From this pool, we focused on peer-reviewed journal articles and eliminated sources including presentations, posters, book reviews, edited book volumes, magazine articles, encyclopedias, and conference proceedings. This yielded 52 total sources.

We then systematically reviewed these articles across seven categories (see Table 1 and Addendum 1 for the full data set). Most of these (categories 1–5) are based upon precedents in related reviews. But to discern some of the qualities that distinguish urban greenways as a distinct type of landscape form and planning practice, we added two categories: "extent" and "landscape setting." Extent refers to the area covered by the greenway under study, and we use the following classification: *urban center* (within a municipal boundary); *metropolitan* (suburban area surrounding an urban center); *rural* (beyond an urban or suburban area); *multi-scalar* (a corridor or network that crosses some combination of urban, metropolitan, or rural areas); and *multiple sites* (studies that examined more than one greenway in different locations).

**Table 1.** Coding sheet.

| # | Review Category | Description | Coding |
|---|---|---|---|
| 1 | Journal | Journal of publication | Text: e.g., Urban Studies |
| 2 | Publication Year | Year of publication | Numerical: e.g., 2006 |
| 3 | Study Location | City where study was conducted | Text: e.g., Sapporo |
| 4 | Research Theme | Main research topic of the paper based on article keywords classified into a modified scheme by James et al. (2009) | Text: experience, management, physicality, valuation |
| 5 | Disciplinary Orientation | Disciplinary orientation of the study | Text: humanities, natural science, social science, interdisciplinary/planning |
| 6 | Extent | Area covered by the greenway under study | Text: urban center, metropolitan, rural, multi-scalar network, multiple sites |
| 7 | Landscape Setting | Predominant landscape setting in which the greenway is located | Text: adaptive reuse, waterfront, active/complete street, new build, multiple settings |

Landscape setting refers to the type of landscape in which the greenway is situated and includes the following classification codes: *adaptive reuse* (greenways developed along spaces that served a previous use such as highways and railways); *waterfront* (greenways that run adjacent to water bodies); *complete street* (greenways that are part of multimodal transit corridors); *new-build* (greenways that are conserved or designed as part of new development); and *multiple settings* (greenways located in two or more of the above). A description of each category and associated classification codes is provided in Table 1.

Categories 1–5 were classified based on deductive (a priori) terms drawn from precedent [35]. However, greenway extent and landscape setting were classified inductively based on terms that emerged in the papers under review. In this case, 20 articles were reviewed and classification terms were established based on this sample. Our original review included both study city and the institutional location of the first author origin; however, there was much overlap between the two and we only report on study location.

In two categories—research theme and disciplinary orientation—we diverted slightly from precedent. Drawing upon Bentsen, Lindholst, and Konijnendijk (2008), we use the term "disciplinary orientation" instead of "type of science," as the former is, in our opinion, a clearer description of the intent and associated coding terms (humanities, natural science, social science, and interdisciplinary/planning) for this review category. In this category, we also use the term "interdisciplinary" instead of "multiple;" and we added "planning" to this classification scheme, as many of the papers qualify as planning studies and many planning studies address both social and ecological concerns.

Based upon the same review precedent, we also adopted the classification scheme of James et al. [40] to depict the main research theme of the paper. Their original scheme included five classification codes: "physicality, experience, valuation, management, and governance." However, we combined "management" and "governance" into a single classification code ("management") because these terms encompass many overlapping ideas, and it was difficult to disambiguate the two. *Physicality* encompasses outcomes related to microclimate, soil, air, and water quality functions and is essentially synonymous with "environmental." *Experience* encompasses people's interaction or contact with green spaces and includes aesthetic, health, and sociocultural dimensions. *Valuation* encompasses links between green space and economic outcomes, and includes topics such as property value and business development. James et al. (2009) also include ecosystem services—human health and well-being benefits of ecosystem functions that are quantified and monetized—in this category; so we classified ecosystem services in both *physicality* and *valuation*. *Management* encompasses planning, design, and governance of urban greenways. In keeping with precedent, we also included *Other* for terms that did not directly classify into a priori categories; however, we removed this classification code from the total count as these terms did not address a research theme. To further minimize risk of misclassification in this category, we systematized the process by using the articles' keywords as the underlying source of data. This had the added benefit of providing quantitative data on the scope of terms associated with each research theme.

For categories requiring little subjective determination (categories 1 to 4), one co-author coded all papers. For categories requiring some subjective determination (categories 5 to 7), both co-authors reviewed all papers and arrived at a shared classification based on definitions and classification codes described above and listed in Table 1.

In addition to these review categories, we also reviewed articles for definitions of urban greenways and applied keywords from these definitions to a word cloud generator. This systematic review provides a foundation for developing a definition and a typology of urban greenways as a subset of greenways writ large. We drew upon Little (1990) [4], Ahern (1995) [1], Hellmund and Smith (2006) [16], and Rupprecht and Byrne (2014) [41] as precedents for the typology which includes descriptions, goals, and examples of five types of urban greenway. Descriptions and goals were based upon review of scholarship and practice. To acknowledge that greenways within each type can be naturalistic or highly constructed, two photographs are included exemplifying each greenway type across this continuum (see Table 7).

## 3. Results

The results of this systematic review are presented below in the same order as the review categories in Table 1.

### 3.1. Journal

As noted in Table 2, *Landscape and Urban Planning* has published the most scholarly articles in the urban greenway literature we reviewed, encompassing 21 out of 52 (40.3%) studies. *Urban Forestry & Urban Greening* has published three articles (5.8%) and six journals—*International Journal of Sustainable Development and World Ecology, Journal of Environmental Planning and Management, Journal of Leisure Research, Journal of Physical Activity & Health, Journal of the American Planning Association*, and *Urban Studies*—have each published two articles (3.8% each). The 16 remaining journals have respectively published one article on urban greenways.

**Table 2.** Number of articles per journal.

| Journal Title | #of Articles | Journal Title | #of Articles |
|---|---|---|---|
| *American Journal of Preventive Medicine* | 1 | *Journal of Outdoor Recreation and Tourism - Research and Planning Management* | 1 |
| *Ecological Engineering* | 1 | *Journal of Physical Activity & Health* | 2 |
| *Ecosystem Services* | 1 | *Journal of the American Planning Association* | 2 |
| *Environment and Behavior* | 1 | *Journal of Urban Planning and Development - ASCE* | 1 |
| *European Journal of Public Health* | 1 | *Landscape and Urban Planning* | 21 |
| *Frontiers of Architectural Research* | 1 | *Landscape Architecture* | 1 |
| *International Journal of Behavioral Nutrition and Physical Activity* | 1 | *Landscape Research* | 1 |
| *International Journal of Environmental Research and Public Health* | 1 | *Professional Geographer* | 1 |
| *International Journal of Sustainable Development and World Ecology* | 2 | *Science of The Total Environment* | 1 |
| *ISPRS International Journal of Geo-Information* | 1 | *Transportation Research Part D: Transport and Environment* | 1 |
| *Journal of Environmental Planning and Management* | 2 | *Urban Forestry & Urban Greening* | 3 |
| *Journal of Leisure Research* | 2 | *Urban Studies* | 2 |

### 3.2. Publication Year

As seen in Figure 1, the first mention of urban greenways in the literature captured in this review was in 1995, coinciding with the first of two Special Issues of *Landscape and Urban Planning* dedicated to this topic [6]. The large spike in 2006 coincides with the second Special Issue of *Landscape and Urban Planning* [8], accounting for six of the 52 total articles. Of note, there has been a relative surge of urban greenway scholarship over the past four years, with four articles published in 2015, five in 2016, five in 2017, and four in 2018.

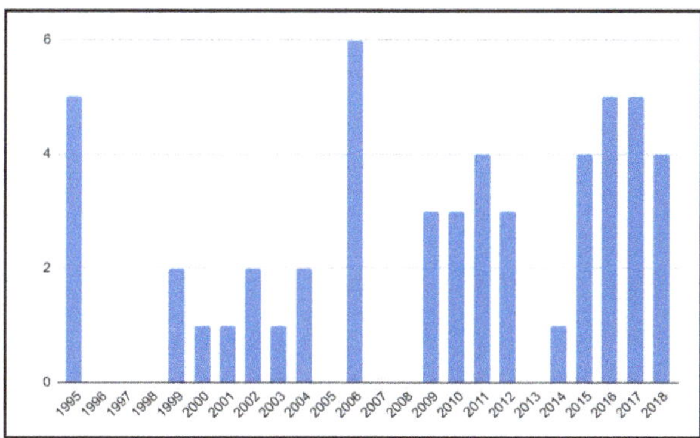

**Figure 1.** Number of publications by year.

### 3.3. Study Location

As noted in Figure 2, most research on urban greenways has been conducted in the United States—especially the eastern region of the country—and parts of Canada, eastern China, and to a lesser extent, Europe. The cities with greenways that have been studied the most are all in the United States, including Indianapolis, IN (5), Atlanta, GA (4), Knoxville, TN (4), Houston, TX (3), Austin, TX

(2), and Chicago, IL (2). Greenways in a few non-U.S. cities have also been studied more than once, including Seoul, South Korea (3), Wuhan, China (3), Shenzhen, China (2), Toronto, Canada (2), and Vancouver, Canada (2). Four articles reference multiple cities in their research: one paper focusing on multiple international greenways [8], one review including ten studies from the U.S. and two from Australia [42], one paper referencing multiple cities across the U.S. [10], and one review paper addressing trails and physical activity that included 49 studies from the U.S., two from Australia, and one from New Zealand [43].

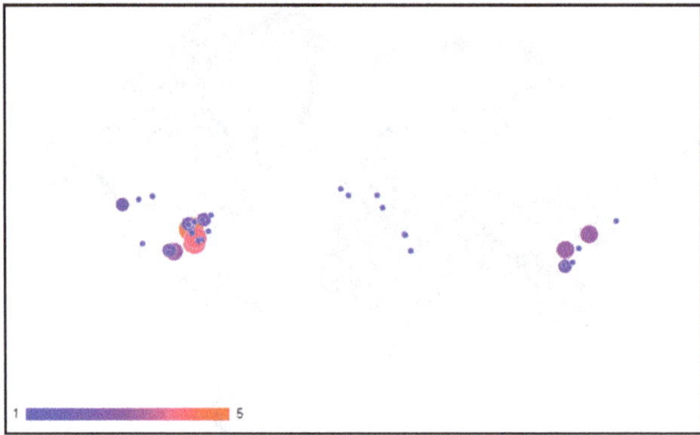

**Figure 2.** Location of greenway study sites by city.

*3.4. Research Theme*

As noted in Table 3, of the 52 articles included in this review, 297 keywords were used to describe the main themes of the research. Of these keywords, 104 did not directly address a research theme (e.g., names of study locations, methods that could be applied to a range of themes). Thus, we removed these terms from the total count, leaving 193 keywords addressing a research theme. In total, 91 keywords (47.2%) focused on human experience, 57 (29.5%) on management, 37 (19.2%) on physicality, and eight (4.1%) on valuation. Two articles did not use keywords: an introduction to a special journal issue [8]; and a longitudinal study on the effects of new urban greenways on transportation energy use and greenhouse gas emissions [44]. For the latter, we added one term to the physicality classification code, making a total of 194 total terms reviewed.

**Table 3.** Research themes by keyword distribution and examples of topic studied.

| Research Theme | Examples of Topics Studied | #of Keywords | Percent |
| --- | --- | --- | --- |
| Experience of greenways | human health, recreation, emotions, perceptions, access, aesthetics, crime, vandalism, | 91 | 46.9% |
| Management of greenways | planning, design, governance, transportation | 57 | 29.4% |
| Physicality of greenways | habitat corridors, biodiversity, landscape ecology | 38 | 19.6% |
| Valuation of greenways | property values, employment density, hedonic analysis | 8 | 4.1% |
| Total | | 194 | 100% |

### 3.5. Disciplinary Orientation

As illustrated in Table 4, of the 52 articles reviewed, 23 (44.2%) are based in the social sciences, 23 (44.2%) reflect an interdisciplinary/planning orientation, and six (11.5%) are based in the natural sciences. None of the articles are based in the humanities.

**Table 4.** Distribution of disciplinary orientation.

| Disciplinary Orientation | Examples of Topics Studied | #of Studies | Percent |
|---|---|---|---|
| Social Science | human health, access, aesthetics, perception, race | 23 | 44.2% |
| Interdisciplinary/Planning | socio-ecological relations, alternative futures | 23 | 44.2% |
| Natural Science | biodiversity, stormwater mgmt., climate change | 6 | 11.5% |
| Humanities | political ecology, discourse, historiography | 0 | 0% |

### 3.6. Extent

As illustrated in Figure 3, of the 52 studies reviewed, 18 (34.6%) studied greenways in urban centers, 11 (21.2%) in the metropolitan region surrounding the urban center, and one in a rural area that modeled the stormwater management capability of greenways in the developing urban-rural fringe [45]. Thirteen (25.0%) of the studies examined multi-scalar greenways that span some combination of urban, metropolitan, and rural. Nine (17.3%) articles discuss multiple greenway case studies occurring at different sites, and thus, spanning different extents.

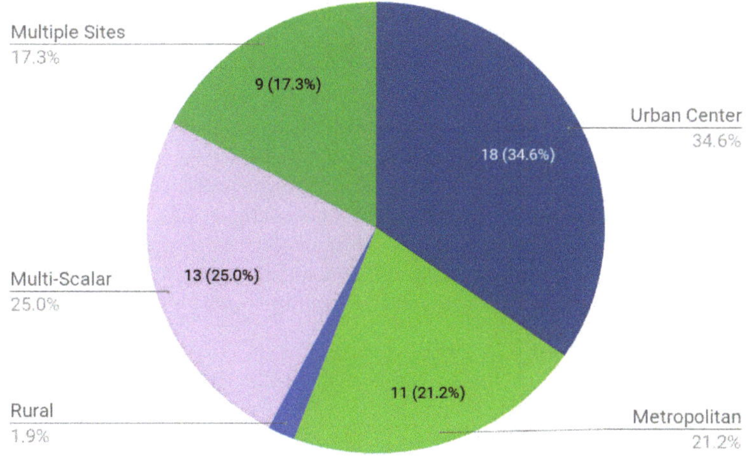


**Figure 3.** Distribution of urban greenway studies by extent.

*3.7. Landscape Setting*

As noted in Figure 4, the majority of articles (60.0%) discuss greenways traversing through more than one landscape setting or multiple greenways in different settings. The next most common setting is adjacent to a water body (22.0%). Adaptive reuse greenways are the focus of 12.0% of studies. Finally, 4.0% of these case studies focus on new on lands that had not previously been developed and one example (2.0%) is part of a complete streets network initiative. Of the 52 articles, two did not clarify setting of the greenway(s) in the study and thus, were omitted from the figure and percentages [46,47].

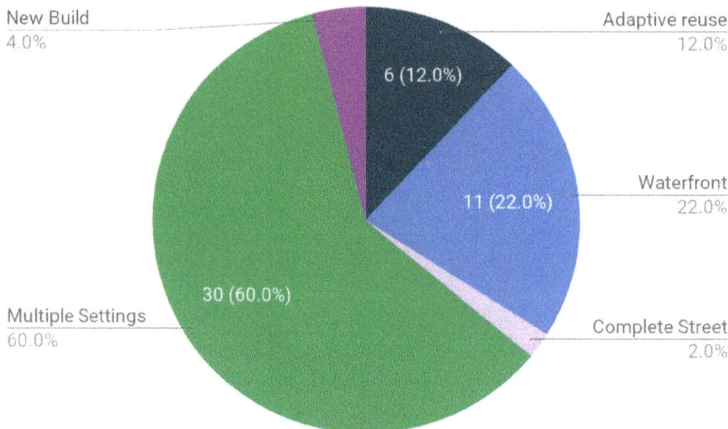


**Figure 4.** Distribution of urban greenway studies by landscape setting.

*3.8. Definitions*

Of the articles reviewed in our sample, seven provided explicit definitions of urban greenways (see Table 5). To distill the most common themes in these definitions, we removed prepositions, conjunctions, particles, irrelevant verbs (e.g., "defined," "be"), adjectives (e.g., "best," "other"), and adverbs (e.g., "often," "generally"), cited sources, and the terms "urban greenway" and "urban greenways." For consistency, we then edited similar words such as "recreation/recreational," "nature/natural," and "public/public realm" to use the same term. In this vein, we also converted "active travel" to "transportation," as the latter is referring to walking and biking in the context at hand, and active travel (or "active transportation") is synonymous with transportation physical activity in public health literature [43,48,49]. The remaining terms were subsequently fed into a word cloud generator, where the frequency of terms is reflected in font size and configuration (see Figure 5).

**Table 5.** Definitions of urban greenways.

| Urban Greenway Definitions | Source |
| --- | --- |
| "Urban greenways which are often designed with multi-use trails that provide opportunities for physical activity, recreation and transportation are defined as places for nature in the city where people can fulfill recreational needs and achieve solitude and retreat without leaving the public realm." | Akpinar 2016 |
| "An urban greenway is generally defined as a linear park and pedestrian-friendly corridor including constructed and natural space." | Jang and Kang 2016 |
| "Almost by definition, urban greenways are places for nature in the city, places where it is sometimes possible to achieve solitude and retreat, without leaving the public realm." | Luymes and Tamminga 1995 |
| "Urban greenways are landscaped and traffic-calmed pathways with a mix of bicycle facilities and other streetscape improvements that link open spaces, parks, public facilities, and neighborhood centers together. Greenways support a variety of active travel uses, including walking, running, bicycling, and skating." | Ngo et al. 2018 |
| The Indianapolis Department of Parks and Recreation defines greenways as: "Multi-use trails intended to connect various neighbourhoods of the city and offer increased alternative pedestrian transportation choices." | Payton and Ottensmann 2015 |
| "For purposes of this paper, we are talking about greenways in urban and urbanizing areas primarily in the USA. Perhaps the best way to find a definition is to look at the two root words, 'green' and 'way'. 'Green' suggests areas that are left vegetated and in most cases appear- or at least strive-to be natural. The word 'way' implies movement, getting from here to there, from point to point. This is the important distinguishing feature of greenways—they are routes of movement-for people, for animals, for seeds, and, often, for water." | Searns 1995 |
| "Urban greenway trails might best be thought of quasi-natural park and open space environments that provide places for daily recreation and alternative transportation options while encouraging positive face to face interaction with other people." | Shafer et al. 2000 |

**Figure 5.** Word cloud of common terms in urban greenway(s) definitions. Developed at https://www.wordclouds.com/.

## 4. Urban Greenway Typology

The assessment of literature described above was complemented with the authors' observation of practice to develop a typology of urban greenways. Table 6 includes descriptions, goals, and examples of five types of urban greenway. Table 7 includes photographs exemplifying each greenway type. We offer two photos for each type to illustrate that urban greenways exist along a naturalistic to constructed continuum. For example, rail-to-trails such as the Promenade Plantée (aka Coulée Verte) in Paris include sections both at street level and up to 30-feet high; while the adjacent landscape on many rail-to-trails can be mostly vegetated.

**Table 6.** Urban greenway typology.

| Type | Description | Goals | Examples |
|---|---|---|---|
| freeway-to-greenway | Adapted from a former highway, elevated or at grade; can include paths, gathering spaces, and programming. | public amenity; cultural resources; community engagement; recreation; nature contact; biodiversity | Cheonggyecheon (Seoul, South Korea); Rose Kennedy Greenway (Boston, USA); Tom McCall Waterfront Park (Portland, Oregon) |
| rail-to-trail | Adapted from a former rail corridor, elevated or at grade; can include small gathering spaces. | active travel; recreation; human health; cultural/historical resources; nature contact | The 606 (Chicago, USA); Promenade Plantée (Paris, France); Capital Crescent Trail (Washington, D.C., USA) |
| waterfront | Adjacent to a water body; hardened, restored, or natural shorelines; can including water access and programmed events. | recreation; waterfront access; riparian protection, stormwater management; cultural resources; environmental stewardship | Guangdong Greenway (China); Schuylkill River Greenway (Philadelphia, USA); Hudson River Greenway (New York, USA) |
| active travel corridor | Pedestrian/cycling transportation corridor adjacent to or in roadway. | active travel; recreation; connectivity; physical activity; human health; reduced greenhouse gas emissions | Comox-Helmcken Greenway (Vancouver, Canada); Emerald Network (Boston, USA) |
| nature trail | Trail through an undeveloped, conserved landscape; generally characterized by a paved or gravel path for pedestrians, hikers, and/or cyclists; can include constructed elements. | recreation; urban containment; wildlife habitat and ecosystem processes; climate change mitigation | Greenville Health System Swamp Rabbit Trail (Greenville County, South Carolina, US); London Green Belt (London, England); Green Wedge Plan (Stockholm, Sweden) |

Note: these urban greenways types are often combined into a network.

**Table 7.** Urban greenway typology: photographic examples.

**Freeway-to-Greenway**

(a). Tom McCall Waterfront Park (Portland, OR, USA) Reproduced with permission from Go4TravelBlog

(b). Cheonggyecheon Greenway (Seoul, South Korea) Reproduced with permission from Jaclynn Seah

**Rail-to-Trail**

(a). Capital Crescent Trail (Washington, D.C., USA) Reproduced with permission from Rails-to-Trails Conservancy and TrailLink.com

(b). Promenade Plantée (Paris, France) Reproduced with permission from Alamy

**Waterfront**

(a). Brooklyn Waterfront Greenway (New York City, USA) Reproduced with permission from Brooklyn Bridge Park Conservancy, photo: ©Etienne Frossard

(b). The Bund along the Huangpu River Walk (Shanghai, China) Reproduced with permission from ©Richard C. Edwards, 2019

**Table 7.** *Cont.*

**Active Travel Corridor**

(a). Minuteman Trail, section of the Emerald Network (Boston, USA) Reproduced with permission from Friends of Lexington Bikeways

(b). Comox-Helmcken Greenway (Vancouver, Canada) Reproduced with permission from Ken Ohrn

**Nature Trail**

(a). Connswater Community Greenway (Belfast, Ireland) Reproduced with permission from the Institution of Civil Engineers

(b). Swamp Rabbit Trail (Greenville, SC, USA) Reproduced with permission from All Trails

Note: The selected images are intended to illustrate that urban greenways exist along a naturalistic (**left** image) to constructed (**right** image) continuum. Additionally, a greenway can include several types and the various types can be combined into a network.

## 5. Discussion

In the ensuing section we discuss the aforementioned results, interpreting patterns in the data with an eye towards implications for scholarship and practice. In addition to addressing each discrete category, this discussion acknowledges synergies between categories.

### 5.1. Journal

As noted in the results, *Landscape & Urban Planning* published over half of urban greenway studies covered in this review. This can be partially explained by the journal having sponsored Special Issues on greenways in 1995 and 2006. Indeed, seven of the 21 studies published by *Landscape and Urban Planning* that were covered in this review, were published in these Special Issues. Considering the multifunctional nature of greenways, and that greenways are a prominent expression of landscape planning practice, this journal is well-suited to the topic at hand. This is reflected in the journal's aims and scope, guided by an underlying premise that "landscape science linked to planning and design can provide mutually supportive outcomes for people and nature" [50]. The prominence of this journal on urban greenways scholarship is also a testament to the legacy of Julius Gy Fábos, Emeritus Professor of

Landscape Architecture at the University of Massachusetts, who co-edited the two aforementioned Special Issues.

*5.2. Publication Year*

As noted, the biggest spikes in urban greenway research coincided with two Special Issues of *Landscape & Urban Planning* dedicated to this topic [6,8]. Special Issues are generally developed when subject experts identify a demand for scholarship in a particular area. High quality Special Issues can, in turn, increase interest in a journal and attract new authors and readers. It should be noted that the recent rise in urban greenway scholarship suggests a broadening disciplinary reach. Of the 18 articles published 2015 to 2018, 14 (77.8%) were published in journals other than *Landscape and Urban Planning*. This temporal assessment also reveals a steady increase in scholarly production. There were eight urban greenway articles published from 1995 to 2000, 12 articles from 2001 to 2006, 13 articles from 2007 to 2012, and 19 articles from 2013 to 2018. It is especially noteworthy that growth in urban greenway scholarship over the past decade has occurred independent of special journal issues dedicated to the topic. This illustrates the degree to which urban greenways have gained traction as an important type of landscape planning scholarship and practice.

*5.3. Study Location*

The geographic distribution of urban greenway scholarship reflects a broader pattern in scholarly production, which tends to be dominated by the U.S. and secondarily, China [51]. Related scholarship in ecological planning and design is also dominated by U.S.-based authors [52]. The lack of many studies in Europe is a bit odd, as greenways have a strong tradition there [10,23,53,54]. Allied scholarship in urban greening and urban forestry also has strong representation in Europe [35]. The lack of urban greenway scholarship in the global south also reflects patterns in scholarly production writ large, and this is a topic of concern. As noted by Ernstson and Sörlin [55], urban environmental research gestures toward frameworks and models that are valid everywhere, and this risks discounting local knowledge and meaning-making. One study in China, for example, showed that in contrast to studies conducted in Western countries, less-educated and low-income respondents visited an urban greenway more frequently than others [56]. As most 21st century urban growth is expected to occur in Africa and Asia [12], greenway scholarship and planning practice will be especially important in these underrepresented areas. Here, research might address the role and potential of greenways in already built-up urban centers, as well as the potential of greenways to shape future urban development.

*5.4. Research Theme*

The literature reviewed in this study found a strong emphasis on research addressing human needs and values as well as those addressing the planning, design, and management of urban greenways. Constituting roughly three quarters of all studies reviewed, this is not surprising: urban areas are, by definition, dense agglomerations of people and cities are inherently complex and contested settings that require nuanced planning and management. What is perhaps a bit surprising, is the relatively limited body of scholarship explicitly addressing environmental issues (classified here as *physicality*). This is especially noteworthy considering the strong ecological foundation that undergirds conceptualization of greenways writ large, as noted in the introduction to this paper [1,5,17,19].

It should be noted, however, that many studies did address environmental concerns but they also included social dimensions, whereby they were classified as interdisciplinary. For example, Larson et al. [57] examined how the public perceives ecosystem services of urban greenways and found that people value cultural benefits, such as social gathering and recreation more than environmental functions. This reinforces the need for thoughtful planning and design and striking a balance between programmed/unprogrammed and naturalistic/hard-scaped spaces.

Other studies embrace a socio-ecological approach that addresses both people and the environment, such as benefits that urban greenways provide for mental health and biodiversity [58], and greenways as strategies for urban sustainability [59–61]. For example, one study found that for residents living near a newly installed greenway, greenhouse gas emissions decreased by 20.9% after the greenway's construction and the change in emissions was attributed to a reduction in vehicle kilometers traveled enabled through provision of high-quality active transportation infrastructure through cycling facilities and other streetscape improvements [44].

It is a bit surprising that there are not more studies addressing the economic dimensions (classified here as *valuation*) of urban greenways. As noted by some studies in the literature reviewed, urban greenways can increase adjacent property values [62,63] and employment density [33]. This can, in turn, be harnessed to finance greenway management through the creation of business improvement districts, exemplified at the Rose Kennedy Greenway in Boston [64]. On the other hand, new green spaces can lead to gentrification and displacement of local residents [65,66]. This tension is ripe terrain for expanded research on urban greenways.

*5.5. Disciplinary Orientation*

The findings of this review category—showing that the vast majority of urban greenway research falls into the domain of social science and interdisciplinary research or planning—dovetail largely with the former review category on research themes. Studies addressing social outcomes, for example, include links between urban greenways and user perceptions [67,68], aesthetic response [69], public access [70,71], physical activity levels [46,72], crime [73], and racial commingling [74].

As noted above, the strong social science orientation of urban greenway scholarship is noteworthy in its differentiation from greenways writ large, which has a strong foundation in environmental science and landscape ecology. Yet, the relatively minor emphasis on natural science should not be interpreted as a lack of attention to environmental concerns. Over 40% of studies in the literature under review adopted an interdisciplinary and/or planning orientation, and this is, in many cases, synonymous with a socio-ecological approach. In urban settings that are built by and for people [28], this is appropriate. Indeed, cities are, in many ways, a classic socio-ecological system where bio-geo-physical elements and processes interact with people and institutions [75,76]. Thus, the results of this review can be seen as heartening evidence that scholarship is responding to the practical realities in which urban greenways are embedded.

The lack of humanities-based scholarship on urban greenways is a noteworthy gap. Humanities scholarship draws upon environmental history and political ecology and often adopts a reflexive position that shines a critical light on the topic at hand. Reflecting upon related research in urban forestry and urban greening, Bentsen et al. [35] suggest that a lack of humanities scholarship can reproduce a meta-narrative that only emphasizes benefits and goods. The same may be true for urban greenways research. For relevant examples of humanities scholarship pertaining to urban greenways that were not captured in this review, see Chung et al. [77] and Safransky et al. [78].

*5.6. Extent*

Reflecting the heterogeneous character of urbanized landscapes, studies in this review category were broadly distributed across urban centers (34.6%), multi-scalar networks (25.0%), and metropolitan areas (21.2%). This suggests that urban greenways scholarship is addressing a range of scales across urbanized areas and it is encouraging to see many studies addressing multiscalar networks that cut across urban and metropolitan extents. For example, Angold et al. [79] found that small mammals may depend on urban greenways extending from the urban center in Birmingham, UK to adjacent boroughs for dispersal. Cook [80] found that an ecological network plan provides modest but important improvement in ecological systems in the Phoenix urban area. Other multi-scalar studies found that trail use can differ by trail segment [81] and by surrounding land use [82]. Of particular relevance to the topic at hand, the latter study found that greenways surrounded by dense residential and mixed land uses, advanced street networks, and large parks were especially important for increasing physical activity. Reflecting Ahern [1], these examples suggest that *networks* of greenways cutting across scales and land uses are important for generating a range of socio-ecological benefits.

*5.7. Landscape Setting*

Many urban greenways are embedded in a range of landscape settings, as revealed through our assessment, which found that well over half (60%) of reviewed studies crossed multiple settings. Another way of interpreting this finding is that urban greenways are doing exactly what they are conceived to do, namely, provide ribbons of green space in landscapes that might otherwise lack green space at all. This is especially true in the complex, heterogeneous fabric of urban areas, where land contestation can make green space provision all the more difficult.

The next most common setting is waterfronts. This is not surprising, as riparian corridors are routinely identified as one of—if not the—most common settings for greenways [1,4,19]. Studies on waterfront greenways address a range of topics, including but not limited to, stream rehabilitation and public access [83], dispersal corridors for invasive trees [84], and links between human perception, safety, and use [85]. One study found that waterfront greenways close to residential zones, employment centers, and key public services such as hospitals and schools increase use [71]. Focusing on the urban-rural fringe, McGuckin and Brown [46] found that stormwater management facilities can be integrated into existing greenways, and if protected during development, can generate a range of socio-ecological benefits.

Reflecting an ongoing movement to reconceive outdated urban landscapes, a handful of studies address one of the boldest and most dramatic "freeways-to-greenways" to date. In Seoul, Korea, the Cheonggyecheon freeway was torn down and replaced by an urban stream and linear park in 2003–2004, making this project both a good example of adaptive reuse and waterfront greenway. Research shows that land value premiums for parcels within the 500 m walkshed of urban greenway entrance points were notably higher than former freeway on-ramps [62]. Related studies also found that employment density increased within a 1.2 km zone surrounding the new urban greenway [33], and that land conversion from single-family residential to commercial was more likely to occur within 1.5 km of greenway pedestrian entrances.

It is worth noting that freeway-to-greenway projects—including early precedents in Portland, Oregon and San Francisco, California—are harbingers of similar projects being explored in other cities [62]. These are complex feats of civil engineering layered with substantial political and economic mobilization. The transformative effect of such projects may foretell increased scholarship on the complex planning, design, and management of such multi-layered landscapes, as well as interconnected effects related to land use change, property value, public access, and equity.

*5.8. Definitions*

As noted in Table 5 and Figure 5, urban greenway definitions highlight certain recurring themes including "transportation," "natural," "public," people," and "places." If we aggregate terms such as "walking," "pedestrian," "bicycling," "skating," "pathways," "routes," "trails," "movement," "traffic-calmed," and "streetscape," it is clear that non-motorized transportation in the form of walking and biking is a dominant idea running across urban greenway definitions. Likewise, terms such as "landscaped," "vegetated," and "natural" connote flora and greenery. In sum, the definitions offered here are largely synonymous with one of four greenway definitions offered by Little [4], p. 1): "any natural or landscaped course for pedestrian or bicycle passage."

One aspect that is, however, missing from this definition is that urban greenways are also "public places," as noted in our review. The notion of "place" is an important, albeit nuanced, idea. The term is often conflated with "space." But space is an abstract term that has no correlation with human experience. In other words, space has no inherent meaning. A place, on the other hand, signifies a space that has social meaning, and this meaning is mediated by human experience [86]. Thus, urban greenways are not merely vegetated corridors for non-vehicular transportation, they are linear public parks that can provide amenities we normally associate with urban parks writ large: places for public gathering; places for nature contact and recreation; and places of civic pride. The Rose Kennedy Greenway, for example, has transformed downtown Boston. Running along the roof of a submerged highway, the greenway includes a series of contemporary parks designed by landscape architects that include public art, food trucks, farmers' markets, fountains, a carousel, and a visitor center for the Harbor Islands, all of which is bound together by vegetated areas. With this in mind, we offer the following definition: "Urban greenways are linear public parks and places that facilitate active travel and recreation in urban areas."

The anthropocentric focus of urban greenway definitions is noteworthy. Indeed, none of the seven definitions that emerged in this review explicitly references non-human environmental processes, organisms, or values. This is quite different from scholarly characterizations and definitions of greenways writ large, where wildlife and ecosystem processes and patterns rooted in landscape ecology figure prominently.

*5.9. Urban Greenways Typology*

The typology of urban greenways described in Section 4 above illuminates some noteworthy traits of urban greenways. In urban centers, greenways are often implemented in highly constructed landscapes that formerly served another purpose. This is exemplified in the Cheonggyecheon freeway-to-greenway, as well as the Rose Kennedy Greenway. Colloquially known as the "Big Dig," this project depressed the Central Artery of Interstate 93—an elevated six-lane highway completed in 1959—and in its place created a 1.5-mile greenway through the heart of the city (see Figure 6). This reflects a late 20th and early 21st century urban parks movement to repurpose and adaptively reuse outdated landscapes such as landfills, elevated rails and highways, and parking lots [32]. It also reflects efforts to redesign the auto-centric landscape that dominated mid-20th century urban planning.

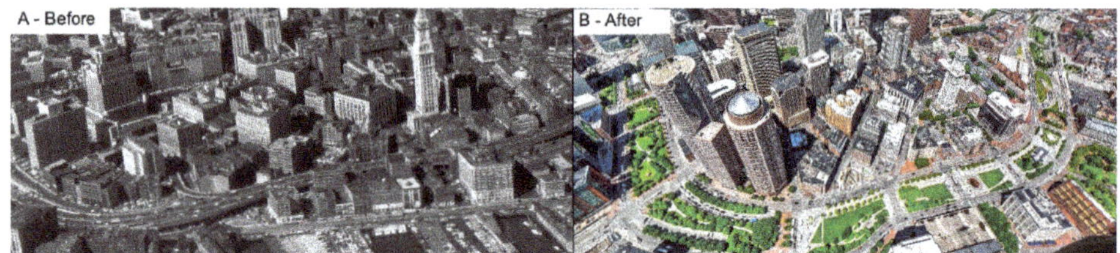


**Figure 6.** (**A**) The Central Artery highway running through the city in 1962. (**B**) Rose Kennedy Greenway in 2017 following depression of the Central Artery. Photo A: Reproduced with permission from the Massachusetts State Archives. Photo B: Reproduced with permission from the Rose Kennedy Greenway Conservancy, photo: ©Kyle Klein

Another contemporary expression of this aspiration is the active travel corridor exemplified in Boston's Emerald Network initiative and the Comox-Helmcken Greenway in Vancouver, Canada (see Figure 7). A unifying theme in these greenways is the redesign of existing streets to accommodate cyclists, including infrastructural interventions, such as: (1) new and upgraded traffic signals; (2) new street paving, concrete medians and curb bulges, catch basins, paint, and signage; (3) new sidewalks, curb ramps, and raised crosswalks; (4) new and upgraded street, sidewalk, and park lighting; and (5) new public realm amenities, such as seating, planting, trees, drinking fountains, and wayfinding features [49].

Similar to the freeway-to-greenway and rail-to-trail examples, these contemporary expressions of urban greenways adaptively reuse existing urban landscapes for new purposes. From a planning perspective, these adaptive reuse types of greenways can be considered an "offensive strategy," in that they introduce new elements in previously disturbed or fragmented landscapes [1]. This resonates with depictions of urban greening as a social practice of organized or semi-organized efforts to introduce, conserve, or maintain outdoor vegetation in urban areas [87,88]. In many cases, adaptive reuse greenways include new plantings and a net increase in flora and biomass. This may not, however, always be the case with active travel corridors, where new plantings are just one of many structural design elements and many stretches may have little if any vegetation.

**Figure 7.** The Comox-Helmcken streetscape before (**A**) and after (**B**) greenway construction. Photo A: Google Street View; Photo B: Reproduced with permission from Paul Krueger, City of Vancouver

This represents a departure from the original conception of greenways, where "green" is synonymous with "nature" or "flora" [4,5,19]. In urban areas, by contrast, "green" coupled with "ways" can signify a broader sustainability and livability agenda. This reflects popular and scholarly discourse on "green cities," which addresses topics including transportation, energy, food systems, and social equity, and climate change in particular [89–92]. The lack of much vegetation in contemporary expressions of greenways as active travel corridors might be perceived as corrupting the greenway construct. It is worth noting, however, that related terms have undergone similar changes. For example, early conceptualization of "green infrastructure" had a strong wildlife conservation orientation that drew upon landscape ecology as a strategy to protect and restore regional hubs and corridors of natural lands/habitat from development [18,93]. The term, however, also became synonymous with low-impact development and stormwater management, where the primary goal is to hold or infiltrate stormwater directly in the ground—often but not always through vegetated systems—instead of channeling runoff into traditional grey infrastructure culverts and pipes that discharge into nearby surface waters [94–96]. Some have even extended the green infrastructure construct to include solar panels, wind turbines, public art installations, and outdoor theaters [97].

Another noteworthy theme of urban greenways is that they exist along a naturalistic to highly constructed continuum. This is clearly the case for freeways-to-greenways, rail-to-trails, and active travel corridors, all of which are new constructions layered upon previous constructions. It is also the case for nature trails, which can include compacted gravel and paved paths as well as constructed bridges. This also applies to waterfront greenways in urban centers, which tend to have hardened shorelines and are often built on fill. But as riverfront greenways extend from urban centers to less densely developed landscapes, the waterfront can assume an increasingly naturalistic condition, as exemplified in the Schuylkill River Greenway in Philadelphia, USA. In some cases, urban greenway planning can include regrading and planting of vegetation to actively restore waterfronts to a naturalistic condition, as exemplified in the Chicago River corridor [98].

## 5.10. Qualifiers

We recognize that there are likely peer reviewed articles addressing urban greenways that this review did not capture. For example, some greenway studies may have been conducted in an urban context but they did not use the term "urban greenway" or "urban greenways" in the title, abstract, or keywords; and there may be studies that were not included in the databases we searched. Thus, this paper should not be read as a comprehensive assessment of any and all peer reviewed studies on greenways in an urban context. On the other hand, urban ecology has shown that the urban context can have different meanings, especially in an urbanizing world where the ecological footprint of cities can extend to global scales [99–101]. Thus, a constructive attribute of this study is that it focuses explicitly on studies addressing landscapes described as urban greenway(s) in the title, abstract, or keywords, and in so doing, provides a focused snapshot of this literature. An additional caveat is that distilling research into discrete categories and sub-coding these categories into discrete classes, can be a subjective and reductionistic act, both of which have inherent problems. Subjective classification is based on the assessors' interpretation, which can introduce unconscious bias and error [102]. Reductionism can, in turn, oversimplify complex relationships [103,104]. With these caveats in mind, categorization and classification can be helpful when seeking to advance understanding of a complex topic or phenomenon.

## 6. Conclusions

This study reports on a systematic assessment of 52 peer-reviewed articles using the term "urban greenway" or "urban greenways" in the title, abstract, or keywords, and covering seven research categories plus definitions. The review finds that there has been an uptick in urban greenway scholarship over the past decade; that urban greenway scholarship and definitions reflect a strong orientation towards human needs and concerns; that many urban greenways adaptively reuse already developed lands; and that the materiality of urban greenways ranges from naturalistic to highly constructed. In urban areas, "green" coupled with "ways" may signify a sustainability and livability agenda that goes beyond vegetation per se. The paper offers a definition of urban greenways and outlines an urban greenways typology that includes: freeway-to-greenway, rail-to-trail, waterfront, active travel corridor, and nature trail. As a subset of greenways writ large, urban greenways reflect an evolving form of landscape planning and design, and an opportunity for associated scholarship and practice.

**Author Contributions:** Conceptualization: T.S.E. Methodology T.S.E., O.S.H. Formal analysis: O.S.H., T.S.E. Investigation: O.S.H., T.S.E. Data curation: O.S.H. Writing-original draft preparation: O.S.H., T.S.E. Writing-review and editing: T.S.E., O.S.H. Visualization: O.S.H. Supervision: T.S.E. Project administration: O.S.H., T.S.E. All authors have read and agree to the published version of the manuscript.

**Funding:** This research received no external funding. However, much of the open access publishing fee was provided by the University of Massachusetts-Amherst.

**Acknowledgments:** We would like to thank Jack Ahern for providing constructive comments on the first author's University of Massachusetts-Amherst Commonwealth Honors College senior thesis, which undergirds this study. We also extend thanks to Maria Håkansson and Tigran Haas in the Department of Urban Planning and Environment at KTH Royal Institute of Technology in Stockholm, Sweden for supporting the corresponding author as a visiting scholar in fall 2019. Additionally, we thank The Rose Kennedy Greenway Conservancy for providing time for the first author to work on this article.

**Conflicts of Interest:** The authors declare no conflict of interest.

## References

1. Ahern, J. Greenways as a Planning Strategy. *Landsc. Urban Plan.* **1995**, *33*, 131–155. [CrossRef]
2. Morris, E.S. New Urban Design Concepts: Greenways and Movement Structure: The Philadelphia Plan. *Archit. Yearb.* **1965**, *11*, 26–40.
3. Whyte, W.H. *The Last Landscape*; University of Pennsylvania Press: Philadelphia, PA, USA, 1968.
4. Little, C.L. *Greenways for America*; Johns Hopkins University Press: Baltimore, MD, USA, 1990.
5. Flink, C.; Searns, R. *Greenways: A Guide to Planning Design and Development*, 1st ed.; Schwarz, L.L., Ed.; Island Press: Washington, DC, USA, 1993.
6. Fabos, J.G.; Ahern, J. Greenways. *Landsc. Urban Plan.* **1995**, *33*, 1–482. [CrossRef]
7. Fabos, J.G.; Ryan, R.L. International Greenway Planning. *Landsc. Urban Plan.* **2004**, *68*, 143–146. [CrossRef]

8. Fabos, J.G.; Ryan, R.L. An Introduction to Greenway Planning around the World. *Landsc. Urban Plan.* **2006**, *76*, 1–6. [CrossRef]

9. Yu, K.; Li, D.; Li, N. The Evolution of Greenways in China. *Landsc. Urban Plan.* **2006**, *76*, 223–239. [CrossRef]

10. Searns, R.M. The Evolution of Greenways as an Adaptive Urban Landscape Form. *Landsc. Urban Plan.* **1995**, *33*, 65–80. [CrossRef]

11. Fabos, J.G. Greenway Planning in the United States: Its Origins and Recent Case Studies. *Landsc. Urban Plan.* **2004**, *68*, 321–342. [CrossRef]

12. Angel, S. *Planet of Cities*; Lincoln Institute of Land Policy: Cambridge, MA, USA, 2012.

13. Mallach, A. *Rebuilding America's Legacy Cities: New Directions for the Industrial Heartland*; Mallach, A., Ed.; The American Assembly, Columbia University: New York, NY, USA, 2012.

14. Schilling, J.; Logan, J. Greening the Rust Belt: A Green Infrastructure Model for Right Sizing America's Shrinking Cities. *J. Am. Plan. Assoc.* **2008**, *74*, 451–466. [CrossRef]

15. Fabos, J.G.; Ahern, J.; Breger, B.; Eisenman, T.S.; Jombach, S.; Kollanyi, L.; Lindhult, M.S.; Ryan, R.L.; Valanszki, I. Adapting to Expanding and Contracting Cities. In Proceedings of the Fábos Conference on Landscape and Greenway Planning, Amherst, MA, USA, 28–30 March 2019; Volume 6.

16. Hellmund, P.C.; Smith, D.S. *Designing Greenways: Sustainable Landscapes for Nature People*; Island Press: Washington, DC, USA, 2006.

17. Forman, R.T.T. *Land Mosaics: The Ecology of Landscapes and Regions*; Cambridge University Press: Cambridge, UK; New York, NY, USA, 1995.

18. Benedict, M.A.; McMahon, E.T. Green Infrastructure: Smart Conservation for the 21st Century. *Renew. Resour. J.* **2002**, *20*, 12–17.

19. Fabos, J.G. Introduction and Overview: The Greenway Movement, Uses and Potentials of Greenways. *Landsc. Urban Plan.* **1995**, *33*, 1–13. [CrossRef]

20. Oxford English Dictionary. Greenway, n. *OED Online*; Oxford University Press: Oxford, UK, 2019.

21. Benedict, M.A.; McMahon, E. *Green Infrastructure. [Electronic Resource]: Linking Landscapes and Communities*; Island Press: Washington, DC, USA, 2006.

22. Lynch, A.J. Is it Good to be Green? Assessing the Ecological Results of County Green Infrastructure Planning. *J. Plan. Educ. Res.* **2016**, *36*, 90–104. [CrossRef]

23. von Haaren, C.; Reich, M. The German Way to Greenways and Habitat Networks. *Landsc. Urban Plan.* **2006**, *76*, 7–22. [CrossRef]

24. McKay, B. An Appalachian Trail: A Project in Regional Planning. In *Architectural Regionalism: Collected Writings on Place, Identity, Modernity*; Canizaro, V.B., Ed.; Princeton Architectural Press: New York, NY, USA, 2006; pp. 225–236.

25. President's Commission of American Outdoors. *American Outdoors: The Legacy, the Challenge*; Island Press: Washington, DC, USA, 1987.

26. Kostof, S. *The City Shaped: Urban Patterns and Meanings Through History*; Thames and Hudson: London, UK, 1991.

27. Costanza, R.; D'Arge, R.; De Groot, R.; Farber, S.; Grasso, M.; Hannon, B.; Limburg, K.; Naeem, S.; O'Neill, R.V.; Paruelo, J.; et al. The Value of the World's Ecosystem Services and Natural Capital. *Nature* **1997**, *387*, 253–260. [CrossRef]

28. Groffman, P.M.; Cavender-Bares, J.; Bettez, N.D.; Grove, J.M.; Hall, S.J.; Heffernan, J.B.; Hobbie, S.E.; Larson, K.L.; Morse, J.L.; Neill, C.; et al. Ecological Homogenization of Urban USA. *Front. Ecol. Environ.* **2014**, *12*, 74–81. [CrossRef]

29. IPBES, (Intergovernmental Science-Policy Platform on Biodiversity and Ecosystem Services). *The Regional Assessment Report on Biodiversity and Ecosystem Services for the Americas: Summary for Policymakers*; IPBES Secretariat: Bonn, Germany, 2018; p. 41. ISBN 978-3-947851-01-0.

30. Wentz, E.A.; York, A.M.; Alberti, M.; Conrow, L.; Fischer, H.; Inostroza, L.; Jantz, C.; Pickett, S.T.A.; Seto, K.C.; Taubenböck, H. Six Fundamental Aspects for Conceptualizing Multidimensional Urban form: A Spatial Mapping Perspective. *Landsc. Urban Plan.* **2018**, *179*, 55–62. [CrossRef]

31. Garvin, A. *Public Parks: The Key to Livable Communities*; W. W. Norton & Company: New York, NY, USA; London, UK, 2011.

32. Harnik, P. *Urban Green: Innovative Parks for Resurgent Cities*; Island Press: Washington, DC, USA, 2010.

33.  Jang, M.; Kang, C.-D. The Effects of Urban Greenways on the Geography of Office Sectors and Employment Density in Seoul, Korea. *Urban Stud.* **2016**, *53*, 1022–1041. [CrossRef]
34.  Rose Kennedy Greenway Conservancy. *Annual Report 2018*; Rose Kennedy Greenway Conservancy: Boston, MA, USA, 2019; p. 12.
35.  Bentsen, P.; Lindholst, A.C.; Konijnendijk, C.C. Reviewing Eight Years of Urban Forestry & Urban Greening: Taking Stock, Looking Ahead. *Urban For. Urban Green.* **2010**, *9*, 273–280.
36.  Galvan, J.L.; Galvan, M.C. *Writing Literature Reviews: A Guide for Students of the Social and Behavioral Sciences*, 7th ed.; Routledge: Taylor and Francis Group: New York, NY, USA; London, UK, 2017.
37.  Boote, D.N.; Beile, P. Scholars Before Researchers: On the Centrality of the Dissertation Literature Review in Research Preparation. *Educ. Res.* **2005**, *34*, 3–15. [CrossRef]
38.  Feak, C.B.; Swales, J.M. *Telling a Research Story: Writing a Literature Review*; University of Michigan Press: Ann Arbor, MI, USA, 2009.
39.  Luederitz, C.; Brink, E.; Gralla, F.; Hermelingmeier, V.; Meyer, M.; Niven, L.; Panzer, L.; Partelow, S.; Rau, A.-L.; Sasaki, R.; et al. A Review of Urban Ecosystem Services: Six Key Challenges for Future Research. *Ecosyst. Serv.* **2015**, *14*, 98–112. [CrossRef]
40.  James, P.; Tzoulas, K.; Adams, M.D.; Barber, A.; Box, J.; Breuste, J.; Elmqvist, T.; Frith, M.; Gordon, C.; Greening, K.L.; et al. Towards an Integrated Understanding of Green Space in the European Built Environment. *Urban For. Urban Green.* **2009**, *8*, 65–75. [CrossRef]
41.  Rupprecht, C.D.D.; Byrne, J.A. Informal Urban Greenspace: A Typology and Trilingual Systematic Review of Its Role for Urban Residents and Trends in the Literature. *Urban For. Urban Green.* **2014**, *13*, 597–611. [CrossRef]
42.  Benton, J.S.; Anderson, J.; Hunter, R.F.; French, D.P. The Effect of Changing the Built Environment on Physical Activity: A Quantitative Review of the Risk of Bias in Natural Experiments. *Int. J. Behav. Nutr. Phys. Act.* **2016**, *13*, 107. [CrossRef] [PubMed]
43.  Starnes, H.A.; Troped, P.J.; Klenosky, D.B.; Doehring, A.M. Trails and Physical Activity: A Review. *J. Phys. Act. Health* **2011**, *8*, 1160–1174. [CrossRef] [PubMed]
44.  Ngo, V.D.; Frank, L.D.; Bigazzi, A.Y. Effects of New Urban Greenways on Transportation Energy Use and Greenhouse Gas Emissions: A Longitudinal Study from Vancouver, Canada. *Transp. Res. Part Transp. Environ.* **2018**, *62*, 715–725. [CrossRef]
45.  McGuckin, C.P.; Brown, R.D. A Landscape Ecological Model for Wildlife Enhancement of Stormwater Management Practices in Urban Greenways. *Landsc. Urban Plan.* **1995**, *33*, 227–246. [CrossRef]
46.  Wolff, D.L.; Fitzhugh, E. The Relationships Between Weather-Related Factors and Daily Outdoor Physical Activity Counts on an Urban Greenway. *Med. Sci. Sports Exerc.* **2010**, *42*, 247–248. [CrossRef]
47.  Quayle, M. Urban Greenways and Public Ways—Realizing Public Ideas in a Fragmented World. *Landsc. Urban Plan.* **1995**, *33*, 461–475. [CrossRef]
48.  Owen, N.; Humpel, N.; Leslie, E.; Bauman, A.; Sallis, J.F. Understanding Environmental Influences on Walking: Review and Research Agenda. *Am. J. Prev. Med.* **2004**, *27*, 67–76. [CrossRef]
49.  Sallis, J.F.; Cervero, R.B.; Ascher, W.; Henderson, K.A.; Kraft, M.K.; Kerr, J. An Ecological Approach to Creating Active Living Communities. *Annu. Rev. Public Health* **2006**, *27*, 297–322. [CrossRef]
50.  Landscape and Urban Planning. Aims & Scope. Available online: https://www.journals.elsevier.com/landscape-and-urban-planning (accessed on 29 October 2019).
51.  SJR, (Scimago Journal & Country Rank). International Science Ranking. Available online: https://www.scimagojr.com/countryrank.php (accessed on 30 October 2019).
52.  Eisenman, T.S. The Ecological Design and Planning Reader. *J. Plan. Educ. Res.* **2016**, *37*, 374–376. [CrossRef]
53.  Jongman, R.H.G.; Külvik, M.; Kristiansen, I. European Ecological Networks and Greenways. *Landsc. Urban Plan.* **2004**, *68*, 305–319. [CrossRef]
54.  De Oliveira, F.L. *Green Wedge Urbanism: History, Theory and Contemporary Practice*; Bloomsbury Academic: New York, NY, USA, 2017.
55.  Lewis, J.; Sawyer, L.; Spirn, A.W.; Ávila, M.; Walker, R.A.; Hoffman, L.M.; Baviskar, A.; Lachmund, J.; Van Sittert, L.; Evans, J.; et al. *Grounding Urban Natures: Histories and Futures of Urban Ecologies*; Urban and Industrial Environments; Ernstson, H., Sörlin, S., Eds.; The MIT Press: Cambridge, MA, USA, 2019.
56.  Chen, Y.; Gu, W.; Liu, T.; Yuan, L.; Zeng, M. Increasing the Use of Urban Greenways in Developing Countries: A Case Study on Wutong Greenway in Shenzhen, China. *Int. J. Environ. Res. Public Health* **2017**, *14*, 554. [CrossRef] [PubMed]

57. Larson, L.R.; Keith, S.J.; Fernandez, M.; Hallo, J.C.; Shafer, C.S.; Jennings, V. Ecosystem Services and Urban Greenways: What's the Public's Perspective? *Ecosyst. Serv.* **2016**, *22*, 111–116. [CrossRef]

58. Cackowski-Campbell, J.; Augustin, S. Research Design Connections: Studies Show the Benefits of Gardens for the Mentally Ill, Reasons for Walking, and How Greenways Help Preserve Urban Biodiversity. *Landsc. Archit.* **2006**, *96*, 72–77.

59. Floress, K.; Baumgart-Getz, A.; Prokopy, L.S.; Janota, J. The Quality of Greenways Planning in Northwest Indiana: A Focus on Sustainability Principles. *J. Environ. Plan. Manag.* **2009**, *52*, 61–78. [CrossRef]

60. Hepcan, Ş.; Kaplan, A.; Özkan, B.; Küçükerbaş, E.V.; Yiğit, E.M.; Türel, H.S. Public Space Networks as a Guide to Sustainable Urban Development and Social Life: A Case Study of Muğla, Turkey. *Int. J. Sustain. Dev. World Ecol.* **2006**, *13*, 375–389. [CrossRef]

61. Lindsey, G. Sustainability and Urban Greenways: Indicators in Indianapolis. *J. Am. Plan. Assoc.* **2003**, *69*, 165–180. [CrossRef]

62. Kang, C.D.; Cervero, R. From Elevated Freeway to Urban Greenway: Land Value Impacts of the CGC Project in Seoul, Korea. *Urban Stud.* **2009**, *46*, 2771–2794. [CrossRef]

63. Payton, S.B.; Ottensmann, J.R. The Implicit Price of Urban Public Parks and Greenways: A Spatial-Contextual Approach. *J. Environ. Plan. Manag.* **2015**, *58*, 495–512. [CrossRef]

64. Lake, R. FAQ: The Greenway Business Improvement District. In *Life on the Greenway*; The Rose Kennedy Greenway Conservancy: Boston, MA, USA, 2018.

65. Betsky, A. The High Line Effect: Are Our New Parks Trojan Horses of Gentrification? *Metropolis* **2016**, *36*, 76–79.

66. Wolch, J.R.; Byrne, J.; Newell, J.P. Urban Green Space, Public Health, and Environmental Justice: The Challenge of Making Cities 'Just Green Enough'. *Landsc. Urban Plan.* **2014**, *125*, 234–244. [CrossRef]

67. Reichhart, T.; Arnberger, A. Exploring the Influence of Speed, Social, Managerial and Physical Factors on Shared Trail Preferences Using a 3D Computer Animated Choice Experiment. *Landsc. Urban Plan.* **2010**, *96*, 1–11. [CrossRef]

68. Shafer, C.S.; Lee, B.K.; Turner, S. A Tale of Three Greenway Trails: User Perceptions Related to Quality of Life. *Landsc. Urban Plan.* **2000**, *49*, 163–178. [CrossRef]

69. Chon, J.; Shafer, C.S. Aesthetic Responses to Urban Greenway Trail Environments. *Landsc. Res.* **2009**, *34*, 83–104. [CrossRef]

70. Akpinar, A. Factors Influencing the Use of Urban Greenways: A Case Study of Aydın, Turkey. Urban For. *Urban Green.* **2016**, *16*, 123–131. [CrossRef]

71. Liu, X.; Zhu, Z.; Jin, L.; Wang, L.; Huang, C. Measuring Patterns and Mechanism of Greenway Use—A Case from Guangzhou, China. *Urban For. Urban Green.* **2018**, *34*, 55–63. [CrossRef]

72. Burchfield, R.A.; Fitzhugh, E.C.; Bassett, D.R. The Association of Trail Use With Weather-Related Factors on an Urban Greenway. *J. Phys. Act. Health* **2012**, *9*, 188–197. [CrossRef]

73. Harris, B.; Larson, L.; Ogletree, S. Different Views From The 606: Examining the Impacts of an Urban Greenway on Crime in Chicago. *Environ. Behav.* **2018**, *50*, 56–85. [CrossRef]

74. Coutts, C.; Miles, R. Greenways as Green Magnets: The Relationship between the Race of Greenway Users and Race in Proximal Neighborhoods. *J. Leis. Res.* **2011**, *43*, 317–333. [CrossRef]

75. Frank, B.; Delano, D.; Caniglia, B.S. Urban Systems: A Socio–Ecological System Perspective. *Sociol. Int. J.* **2017**, *1*, 1–8. [CrossRef]

76. Grove, J.M.; Burch, W.R. A Social Ecology Approach and Applications of Urban Ecosystem and Landscape Analyses: A Case Study of Baltimore, Maryland. *Urban Ecosyst.* **1997**, *1*, 259–275. [CrossRef]

77. Chung, C.K.L.; Zhang, F.; Wu, F. Negotiating Green Space with Landed Interests: The Urban Political Ecology of Greenway in the Pearl River Delta, China. *Antipode* **2018**, *50*, 891–909. [CrossRef]

78. Safransky, S. Greening the Urban Frontier: Race, Property, and Resettlement in Detroit. *Geoforum* **2014**, *56*, 237–248. [CrossRef]

79. Angold, P.G.; Sadler, J.P.; Hill, M.O.; Pullin, A.; Rushton, S.; Austin, K.; Small, E.; Wood, B.; Wadsworth, R.; Sanderson, R.; et al. Biodiversity in Urban Habitat Patches. *Sci. Total Environ.* **2006**, *360*, 196–204. [CrossRef]

80. Cook, E.A. Landscape Structure Indices for Assessing Urban Ecological Networks. *Landsc. Urban Plan.* **2002**, *58*, 269–280. [CrossRef]

81. Lindsey, G. Use of Urban Greenways: Insights from Indianapolis. *Landsc. Urban Plan.* **1999**, *45*, 145–157. [CrossRef]

82. Liu, K.; Siu, K.W.M.; Gong, X.Y.; Gao, Y.; Lu, D. Where Do Networks Really Work? The Effects of the Shenzhen Greenway Network on Supporting Physical Activities. *Landsc. Urban Plan.* **2016**, *152*, 49–58. [CrossRef]

83. Che, Y.; Yang, K.; Chen, T.; Xu, Q. Assessing a Riverfront Rehabilitation Project Using the Comprehensive Index of Public Accessibility. *Ecol. Eng.* **2012**, *40*, 80–87. [CrossRef]

84. Säumel, I.; Kowarik, I. Urban Rivers as Dispersal Corridors for Primarily Wind-Dispersed Invasive Tree Species. *Landsc. Urban Plan.* **2010**, *94*, 244–249. [CrossRef]

85. Asakawa, S.; Yoshida, K.; Yabe, K. Perceptions of Urban Stream Corridors within the Greenway System of Sapporo, Japan. *Landsc. Urban Plan.* **2004**, *68*, 167–182. [CrossRef]

86. Tuan, Y.-F. *Space and Place: The Persepective of Experience*; University of Minnesota Press: Minneapolis, MN, USA, 1977.

87. Eisenman, T.S. Greening Cities in an Urbanizing Age: The Human Health Bases in the Nineteenth and Early Twenti-First Centuries. *Chang. Time* **2016**, *6*, 216–246. [CrossRef]

88. Feng, Y.; Tan, P.Y. Imperatives for Greening Cities: A Historical Perspective. In *Greening Cities: Forms & Functions*; Tan, P.Y., Jim, C.Y., Eds.; Springer: Singapore, 2017; pp. 41–70.

89. Campbell, S. Green Cities, Growing Cities, Just Cities?: Urban Planning and the Contradictions of Sustainable Development. *J. Am. Plan. Assoc.* **1996**, *62*, 296–312. [CrossRef]

90. Beatley, T. *Green Cities of Europe: Global Lessons on Green Urbanism*; Kindle edition; Beatley, T., Ed.; Island Press: Washington, DC, USA, 2012.

91. Earth Day Network. Green Cities. Available online: https://www.earthday.org/campaigns/green-cities/ (accessed on 1 December 2019).

92. Lindfield, M.; Steinberg, F. (Eds.) *Green Cities*; Urban Development Series; Asian Development Bank: Manila, Philippines, 2012.

93. Schrijnen, P.M. Infrastructure Networks and Red-Green Patterns in City Regions. *Landsc. Urban Plan.* **2000**, *48*, 191–204. [CrossRef]

94. Prince George's County, MD. *Low-Impact Development Design Strategies*; Prince George's County, MD; Department of Environmental Resources: Largo, MD, USA, 1999.

95. City of Philadelphia. *Green City, Clean Waters: The City of Philadelphia's Plan for Combined Sewer Overflow Control*; Philadelphia Water Department: Philadelphia, PA, USA, 2011; p. 46.

96. City of New York, (Department of Environmental Protection). *NYC Green Infrastructure Plan: 2011 Update*; The City of New York: New York, NY, USA, 2012.

97. Rottle, N.D.; Maryman, B. Envisioning a City's Green Infrastructure for Community Livability. In *Community Livability: Issues and Approaches to Sustaining Well-being of People and Communities*; Wagner, F., Caves, R., Eds.; Routledge: Taylor and Francis Group: London, UK, 2012.

98. Gobster, P.H.; Westphal, L.M. The Human Dimensions of Urban Greenways: Planning for Recreation and Related Experiences. *Landsc. Urban Plan.* **2004**, *68*, 147–165. [CrossRef]

99. MacGregor-Fors, I. Misconceptions or Misunderstandings? On the Standardization of Basic Terms and Definitions in Urban Ecology. *Landsc. Urban Plan.* **2011**, *100*, 347–349. [CrossRef]

100. McIntyre, N.E.; Knowles-Yánez, K.; Hope, D. Urban Ecology as an Interdisciplinary Field: Differences in the Use of "Urban" Between the Social and Natural Sciences. In *Urban Ecology*; Marzluff, J.M., Shulenberger, E., Endlicher, W., Alberti, M., Bradley, G., Ryan, C., Simon, U., ZumBrunnen, C., Eds.; Springer: Boston, MA, USA, 2008; pp. 49–65.

101. Pickett, S.T.A.; Cadenasso, M.L.; Childers, D.L.; McDonnell, M.J.; Zhou, W. Evolution and Future of Urban Ecological Science: Ecology in, of, and for the City. *Ecosyst. Health Sustain.* **2016**, *2*, e01229. [CrossRef]

102. Takacs, D. How Does Positionality Bias Your Epistemology? *NEA High. Educ. J.* **2003**, 27–38.

103. Eisenman, T.S.; Churkina, G.; Jariwala, S.P.; Kumar, P.; Lovasi, G.S.; Pataki, D.E.; Weinberger, K.R.; Whitlow, T.H. Urban Trees, Air Quality, and Asthma: An Interdisciplinary Review. *Landsc. Urban Plan.* **2019**, *187*, 47–59. [CrossRef]

104. Menzies, P. Against Causal Reductionism. *Mind* **1988**, *97*, 551–574. [CrossRef]

 *land*

Article

# Amman (*City of Waters*); Policy, Land Use, and Character Changes

**Anne A. Gharaibeh [1,*], Esra'a M. Al.Zu'bi [1] and Lama B. Abuhassan [2]**

[1]  Department of City Planning and Design, College of Architecture and Design, Jordan University of Science and Technology, Irbid 22110, Jordan; esraaal.zubi@yahoo.com

[2]  Department of Architecture, Faculty of Architecture and Design, Petra University, Amman 11196, Jordan; lama.abuhassan@uop.edu.jo

*  Correspondence: dr-anne@just.edu.jo; Tel.: +962-7201000-26747

Received: 30 October 2019; Accepted: 6 December 2019; Published: 15 December 2019

**Abstract:** The character of Amman, Jordan, as the *"City of Waters"*—referring to the abundance of water flowing in its known stream—has faded away because of the municipal policy to cover the stream in the 1960s which gradually changed the ecological character. This paper traces and explores the impacts of stream-coverage policy on the city character, morphology and land use changes. The purpose is to understand how an engineered problem-solving policy changed physical and perceptive factors and affected the character of the city. It also explores future attitudes towards reversing the non-nature-friendly conditions. The methods depend upon monitoring morphological changes in aerial photographs and in land use maps from municipality archives, conducting interviews with the elderly who witnessed change, one-to-one questionnaires with stakeholders and online questionnaires with residents and visitors. The results show that covering the stream is depriving the city of its historical/ecological character. The policy failed to promote affluent business, to mitigate flood impacts, or to decrease traffic congestion in the Central Business District (CBD). Most age groups believe the stream can improve the image and economy, despite the fact of their unawareness of its historical presence. In conclusion, engineered problem-solving should not stay in the hands of decision makers (technocrats) alone, but rather be considered with the public, sustainable character experts, and ecologists.

**Keywords:** stream coverage; greenway; city character change; land use change; development policy; green corridor; green infrastructure

## 1. Introduction

Throughout history, humans have altered streams, corridors, rivers and other hydro-systems to serve urban areas. Waterways as the kernel for urban settlements and development also affect the artistic quality of urban forms, as well as the functionality and size of cities. In many cases, rapid urbanization and growing cities increased demands on services and pressure on the growth of infrastructure, jeopardizing waterways and ecological characteristics in general [1–3]. While serving the growing demand for transportation systems, many ecosystems are slowly vanishing from urban areas, and at the end are affecting the city character and sustainability [2,3]. Changes in land cover and land use, biodiversity, ecosystems and hydro-systems are also affecting the climate, both locally and regionally [4,5]. Impervious surfaces of asphalt and concrete will reduce areas covered by water and vegetation, decrease evapotranspiration, increase both runoff and the surfaces absorbing solar energy, and increase temperature of the city especially in semiarid environments [3].

In order to gain more urban surfaces, some policies adopt the coverage of natural waterways impacting ecology, health, and socioeconomic factors, causing serious flood risk due to the increased likelihood of blockages, changing the city character and reducing recreational value [6].

This study is a longitudinal temporal study of the Amman Stream corridor, focusing on the life-course developments, the land use and city character changes. It concentrates on exploring the impact of stream coverage and its conversion to a culvert in the downtown. Implementing this policy took the municipality a number of years (1960s–1990s). As a result, the city lost its ecological green corridor, with the flowing stream diminishing any connotations to its ancient character as the *"City of Waters"* [7,8]. This study will clarify the vision of the city, the nature of its morphology, the changes in land use and the perception of the city. In addition, it will explore opinions towards the stream revival issue. It may encourage public commitment to a city character reversal. The following sections will cover the history of the stream, policies associated with developing Amman and the impact of covering the stream on the city.

### 1.1. History of the City of Waters

Amman is known among historians for its many sources of water from streams, springs, artesian wells and reservoirs along its valleys. It is also described by the Torah as *"The royal city"* i.e. the capital, or that *"The city of waters"* [7]. The name Amman came from the time when it was the capital for the Ammonite kingdom (1200 BC). Later, the Greco-Romans made Amman one of the Decapolis league and named it Philadelphia. Lower Amman was located on the banks of the stream where people built their houses from stones and mud, and where they practiced their daily lives, relying on trade, agriculture and grazing. Upper Amman was atop the citadel mountain, which rose about 130 meters from the stream of Amman. Ponds and wells collected water and stored it for use during the siege and in the summer season.

Overall, the basic promoter for urban agglomerations was the stream running in its corridor. A collection of natural springs from Ras Al-Ein (formerly known as Wadi Abdoun); and the rainwater runoff from the western mountains maintained the stream flow. The fresh waters supported agricultural lands and provided the basic everyday needs of the settlers. Amman was famous in the Umayyad era (636–750 AD) for the cultivation of grain and animal husbandry [9–11]. The manifestations of prosperity and advancement of the city at the Umayyad era are clear when we know that the money mintage was on its land. It also sustained itself will into 1347 AD. The importance of Amman as a trading center and an exchange market has been further enhanced by the presence of several flourmills (run mills), turned by the flowing stream waters.

The historical continuity and population sequence in Amman witnessed a long recession caused by earthquakes and epidemics, which lasted several centuries (1347 AD to 1878 AD). In the late nineteenth century, Circassians fleeing Russian genocide came and settled in Amman, where it began to grow and flourish again. Therefore, the history of modern Amman dates back just to the late nineteenth century.

Circassians started coming to Amman as early as 1878, and landed in the Ras Al-Ein stream corridor, around the ancient Omari Mosque (Umayyad Mosque, later known as the Husseini Mosque). By 1895, Amman became full of life, traffic and urbanization (along with agricultural activities and handicrafts) [12,13].

The population of the city of Amman in the 1920s was estimated between 3000 and 5000, indicating the small size of the city (See Figure 1 for the chronological photo order). In 1921 occurred the establishment of Amman as the capital of Transjordan [9–11]. In 1930, its population increased to 10,000, due to the attraction of many people seeking work in the capital, in addition to the internal and external migrations [11]. The population growth in Amman has been neither gradual nor natural; the city has witnessed population mutations in 1948, 1967, 1976 and 1990 as a result of the political conditions in neighboring countries. Amman transformed from a small village with a population less than 100,000 into a city with three million inhabitants by the end of 1987 (and over four million now) [14].

**Figure 1.** Timeline photographs of Amman downtown 1900–1985 AD [15].

The huge demographic mutations and migration caused an increased demand for infrastructure, transportation and urban expansion. Decision makers considered that pressures posed negative impacts that needed quick mitigation measures. Focusing on the problem caused misfortunate decisions that jeopardized the natural resources of the city in order to absorb the increasing population densities.

*1.2. Developing Amman; Plans and Policies*

Amman has witnessed different development plans over the years 1955, 1968 and 1970 by many authorities and organizations. These plans were a reflection of the frequent demographic and infrastructure changes in Amman. The plans focused on the city kernel by the stream, and its surrounding urban areas, and proposed changes to the stream based on the plan objectives.

In 1955, the United Nations (UN) field planners and consultants, King and Lock, prepared the first comprehensive development plan for the city of Amman (Amman Municipality archives, 2019) (Figure 2). This plan aimed at achieving objectives for housing and employment sectors according to the big flow of refugees after the Palestinian–Israeli first war in 1948. The plan aimed to preserve and enhance the nature through the establishment of self-contained mountain neighborhoods. It embraced the concept of "green fingers", which fringed from the stream corridors and into the heart of the city, creating public open spaces. The plan proposed a "central park" in the city center, which included the municipal building, library, theater and arts gallery [9]. Clearly, this plan followed Ebenezer Howard's Garden City ideas, and the British town planning innovations of the 1640s [9]. Based on later plans by the municipality, this idea was not implemented or developed any further in terms of land use management or spatial planning processes. The green fingers and the ecological city core did not see the light. Piecemeal additions that responded to the boom cycles of building activity replaced the 1955 visionary plan.

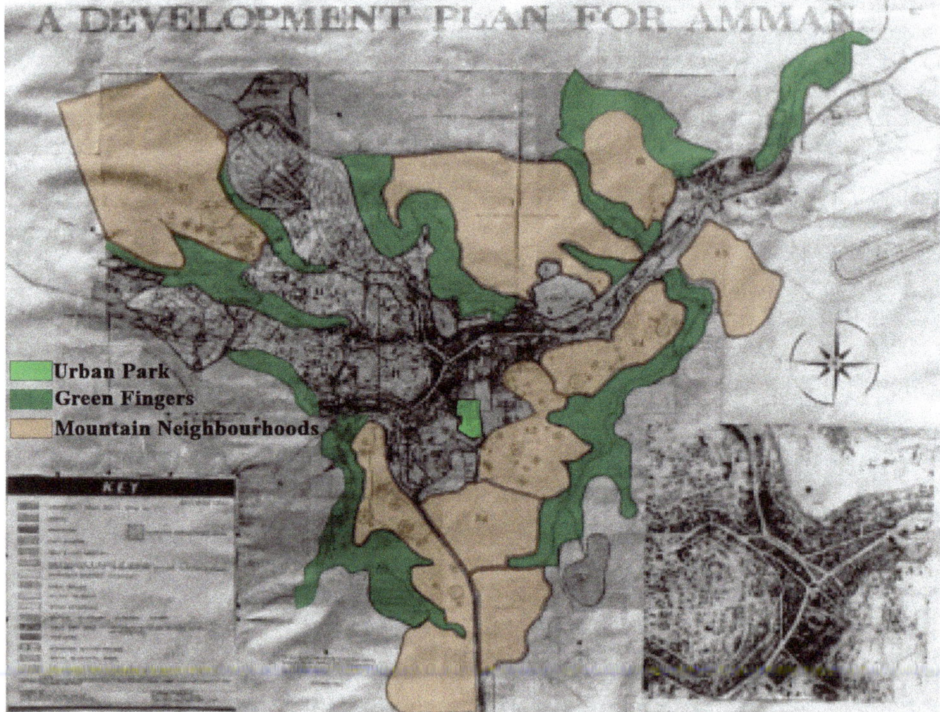



**Figure 2.** Development plan, 1955 source: [9]. For clarity purposes, the authors colored the green fingers and future mountaintops that are planned to become future urban areas.

In the 1960s, the Planning Department was established in the Jordan Development Board (JDB), where they affirmed the need to improve the agricultural and tourism sectors, and to achieve evolution and development in both sectors through a Seven Year Program for the Economic Development of Jordan (1964–1970). However, according to the decline of the Jordanian economic situation during the Israeli occupation of the West Bank in 1967, the Civic Center Development Plan (CCDP) in 1968 has been prepared to focus mainly on tourism 148 redevelopment of the CBD of Amman. The proposed plan consisted of a linear park along with other administrative, commercial, recreational and service land uses [9].

The growth witnessed unplanned population mutations (1967–1970) because of neighboring political factors, which resulted in urban expansion and urban mutations in the city. These urban mutations increased the need for infrastructure, the need for new land uses and the need for urban expansion at the expense of agricultural lands and natural resources. This urban expansion led to jeopardizing the stream and the natural resources associated with it which mediated the city, giving it a unique character throughout history. Because of this transformation, the uses that existed on both sides of the stream had changed, causing a character transformation.

### 1.3. The Coverage of the Stream

The radical changes began when implementing the 1967 initiative plan, in which the government worked on a project for land acquisition adjacent to the stream. The project converted the stream into a culvert whose top became Quraysh Street. In addition, the initiative converted the residential land use to commercial and central business uses, aiming to improve the situation of the downtown and make it the commercial center of the capital. After the implementation of the plan (early 1970s), the radical changes began to take place (see Figure 3).

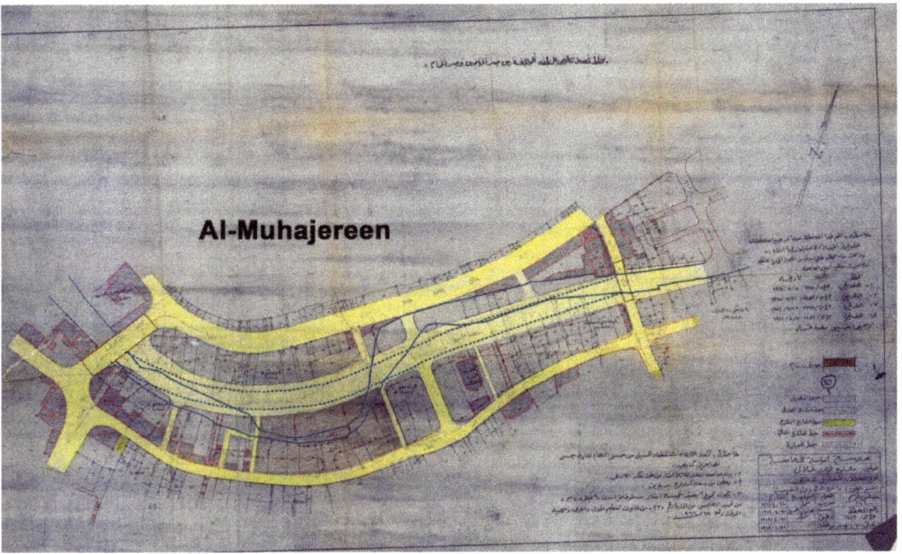


**Figure 3.** Blueprint of the planning modifications of 1967, excerpted from Greater Amman Municipality archives, recolored by the authors.

The municipality's policy of covering the stream continued even after the 1970s. The target area was from the Roman amphitheater towards Ein Ghazal, as demonstrated through the 1971 modification plan prepared by the municipality to redistribute land and to change the uses to commercial land use (Supplementary Materials).

In a report conducted by the Al-Rai newspaper about the coverage of the stream of Amman, Al-Azra'e [11] summarized the several threats which The Greater Amman Municipality hoped to overcome by covering the stream:

1.  The poor health conditions caused by the proliferation of insects such as mosquitoes and flies, despite repeated sterilization campaigns.
2.  The poor social and economic conditions of the residents on both sides of the stream, who are more than 50 thousand citizens.

3.  The winter flooding that periodically turns the stream into an energetic river that sweeps away all around it, including meager fences and old houses on both sides of the stream.
4.  The traffic problems by creating parallel streets and one-way traffic. They realized that if the covered stream was turned into a street it would relieve the pressure from main streets like King Talal Street.

Accordingly, the coverage of the stream stretched in phases to 10km; extended between Ras Al-Ein and Ein Ghazal (Supplementary Materials). The plan included organizing the area on the stream between the former Amman Municipality building and Raghadan Bridge, which will turn into a square where the public library, exhibitions, theaters and other cultural facilities will be built. The plan also included the rehabilitation of the area through the establishment of commercial buildings. The Municipality proposed future land use plans that included central business and central commercial uses to enhance the commercial character of the downtown (Municipality archives). The seventies' policy aimed to solve the problems of traffic, winter flash floods and pollution, as well as to develop the downtown to become the commercial center of Amman by covering the stream and converting it to a culvert under Quraysh Street.

*1.4. The Current Situation in the Downtown*

The stream before the coverage used to flow in a valley surrounded by mountains, where the rainwater assembled and discharged in the stream corridor. At some times it flowed with ease, at other times it flooded the banks and some of the surrounding functions. One of the objectives of the covered culvert was to reduce the problem of recurrent flooding of the stream in the winter season, creating a more secure downtown [16]. Since the establishment of the culvert in the 1960s, the rainwater was flowing into the stream culvert and sometimes raising its level, threatening residents living near the stream (Figure 4A). In the winter of 2015, the downtown received high levels of rainfall, causing the loss of four souls in the flash floods [16].

In February 2019, the downtown witnessed heavy rainfall again, which led to flash floods [17]. These floods led to landslides in the area of Jabal Al Jofah. The collapse damaged a number of vehicles in the near area. In addition, the floods caused the closure of the entire Quraysh Street, risking the safety of citizens, causing car drowning and shop flooding on both sides of the street (Figure 4B).

The Greater Amman Municipality (GAM) has declared that the reason for the floods on Quraysh Street at the downtown was due to the blockage in the manholes because of soil erosion in Jabal Al Jofah, bringing amounts of rainwater mixed with soils and debris. The president of the Association of Food Merchants estimated the losses in the downtown in millions of Jordanian dinars, caused by floods that entered the shops and spoiled their goods [17]. The Amman Chamber of Commerce estimated the losses suffered by 60 shops to be more than 2.8 million USD. They explained that the reason for these losses was the poor infrastructure in the downtown, which could not absorb the amounts of rainwater, which in turn destroyed the existing infrastructure [18]. The flooding could have happened even if the stream was uncovered, but the type of damage would be different. It will affect agricultural and recreational lands by its banks. In either way, due to climate change matters, Amman is receiving more rain in the future.

The 1960s policy of covering the stream was focused on creating more streets for traffic. Despite the invention of Quraysh Street atop the stream along with King Talal Street in the CBD, these streets are not coping with the traffic demand. The downtown still suffers from traffic congestion for many hours throughout a normal working day. The capacity of the infrastructure can no longer meet the growing population with the high density of private vehicles [15–19].

**Figure 4.** Flash floods in the downtown, row **A**: flash floods in the last century, row **B**: flash floods on the 28th of February 2019 (Source of row **A**: [20], Source of row **B**: [17].

## 2. Literature Review

This section will discuss the importance of green infrastructures in attracting other uses. It will present four case studies of policies that turned watercourses into green infrastructures. It discusses land use change, the use of urban greenways and green corridors, threats to character change and the public perception of city character.

Some studies have shown the positive impact of existing natural and environmental facilities in achieving economic land use development [13,21–23]. Green amenities and urban greenways played a critical role in attracting communities and land use development. In the United States, residents prefer to live near parks and greenways, thus land use development has resulted from people's mobility to green amenities, such as urban greenways [24–26]. In Egypt, the local residents benefited from the existence of the urban greenway in linking their neighborhoods, promoting education and recreation, as well as preserving the natural environment and raising the awareness about its importance. In addition, greenways attract business, commercial and cultural uses when they intersect with residential areas [26].

People have strong feelings towards their natural environments. This social need is subject to change based on municipal decisions. Some policies have succeeded in guiding the compass towards environmentally and urbanely neglected areas, through making these areas safe and attractive urban areas that entice investments and population alike. The next sections will introduce four case studies; Cheonggyecheon, Seoul, South Korea; Los Angeles River, LA, USA; San Antonio Riverwalk, TX, USA; and Bartin River, Turkey.

In South Korea, the restoration of Cheonggyecheon stream came after more than 30 years of covering up the watercourse with concrete and an expressway. The stream was first converted into an elevated freeway project, which caused years of degraded, poor environmental conditions. Lastly, the Seoul mayor initiated a project to remove the elevated freeway and to restore the stream, providing an urban greenway [27,28]. This restoration gained the city economic and development revenues and created land use change. The restoration led to conversion of the residential land use into commercial and mixed land uses surrounding the greenway, while the residential uses shifted to farther distance with higher densities.

The Los Angeles River has witnessed another successful example in reviving the city character. This Los Angeles River was originally an alluvial river, crossing floodplain areas on which Los Angeles and other surrounding towns were built [29–31]. Until 1825, the river discharged into the sea. Later, the river way to the sea became muddy and blocked the river flow.

The waters flooded and spread all over the land adjacent to the river and filled the lowlands, creating lakes, swamps and ponds [32]. In 1913, the Los Angeles Aqueduct was opened. The river was the main source of water at that time, despite the fact that the water was shallow for most of the time except winter [29]. In the 1930s, frequent floods devastated the river, therefore, the mayor of Los Angeles, Frank L. Shaw (1877–1958), worked in cooperation with the US Army Corps of Engineers (USACE) on several measures to control and reduce floods. They worked on a project to cover the riverbed and its banks with concrete, and turned it into a canal intermediated with a small flow of water [31]. Channelizing the river led to the drying up of the river for nine months of the year. The conditions repeated until the 1950s, when the water was re-pumped into the canal with the industrial and residential discharge of gray water. In 1986, the Friends of the Los Angeles River (FoLAR) organization was established to revive the river and to restore habitat and public access to the river [30–33]. Later, the nonprofit group River LA (formerly, Los Angeles River Revitalization Corporation) announced in 2013 the goal of completing about 82km of greenway and bike path along the river by the end of 2023. This route is expected to be the main hub of a Linear park, in addition to providing an alternative transportation route through the city [33]. The USACE conducted a feasibility study on the river restoration project. They recommended a plan worth $453 million for flood protection and ecosystem restoration for 600 acres [34].

The San Antonio Riverwalk presents one of the most successful walkway networks. The Riverwalk, located along the San Antonio River in Texas, is considered an important part of the urban fabric of the city. It is an attractive destination to local residents and tourists due to the lining of shops, restaurants and historical missions integrated with nature on both riverbanks [35,36].

A catastrophic flooding of the San Antonio River killed 50 people in 1921. To reduce the risk of flooding, a flood control plan was initiated including the construction of a dam at the source of San Antonio River, and the creation of a storm sewer by paving the bend of the river [37,38]. In 1926, the San Antonio Conservation Society protested against the plan of paving the sewer and prevented the implementation of the plan. In 1929, architect Robert Hugman (1902–1980), presented a plan to turn it into a Riverwalk surrounded by commercial development. Despite the much interest in the development and construction of the site, the architect and later mayor Mr. Jack White played a key role in enabling the San Antonio River Beautification Project through the idea of issuing bonds to raise funds to support the project in 1938 [37–39]. Congressman Maury Maverick, Mayor C. K. Quin (in office 1933-1942) and a group of citizens, headed by White began the site development [37–39].

During the successive decades, there have been improvements and expansions along the Riverwalk. The development incorporated several stages of expansions and improvements: 1968, 1981, 1988, 2009 and 2011. Now improvements and ecological controls include the long-distance walkways, cycling trails and rowing trails, which allowed accessible touristic experiences [37–39].

In Turkey, a similar case found that the Bartın River pertained a significant potential for creating a preserved land use balance to promote future urban development. It was foreseen to support the environmental health and aesthetic character in the city. The preservation will mitigate food and flood risks on the river corridor [40].

Policies turned Cheonggyecheon in Seoul, the San Antonio River and the Los Angeles River into effective greenways and green corridors that improved the green infrastructure of their cities. Restoring the Seoul stream into a greenway was considered as an important policy to reintroduce nature to the city which led to a reduction of the urban heat island effect by as much as eight degrees centigrade, in comparison to nearby paved roadway conditions according to summertime measurements [41]. To promote a more ecofriendly urban design, it brought improvements in water quality as well as the creation of a natural habitat [42]. In addition, it restored the history and culture of the region and revitalized Seoul's economy. As traffic decreased, air quality improved in the created linear park along the stream (greenway) with landscaping, good walking facilities and plenty of street furniture. It slowly became an attractive destination for both residents and businesses, which led to land use change and economic development [27,28].

One of the most important ecological issues highlighted over the last decades is turning green corridors into culverts. Policies now focus on preventing the coverage of natural waterways by promoting the transformation of these structures, and restoring urban waterways back into more natural courses [6,43]. Many countries of the world have taken the approach of restoration of urban waterways as an opportunity to reduce the shortage of natural open spaces and enhance the wellbeing of the local communities through affording linear recreational areas within the cities [44].

Urban greenways have played an important role in the promotion and the development of urban and suburban environments. They provided the most needed natural corridors inside the urban areas, mitigated the loss of natural spaces which are affected by the urban expansion and constituted the primary resistant form to the built environment [26,45,46]. Greenways approach aims to protect nature which balances between the conservation and the growth through creating livable environments and maintaining open spaces. They help maintain biological diversity, protect water resources, conserve soils, improve air quality, reduce pollution, support recreation, enhance community and cultural cohesion, provide species migration routes during climate or seasonal change, manage water runoff, contribute to aesthetic qualities, enhance economic values, diversify recreational opportunities through activities and protect natural character [47–51]. Greenways could thus be considered lines of opportunity; they contribute to many ecological and societal values, reviving cities and preserving character [45].

The 'town character' or 'city character' terms often referred to the sum of the distinctive features and elements that make the town unique. The character is associated with environmental features, and with the response to the environmental and social aspects of the town, by which people judge the suitability of changes in their surrounding environment [52]. The concept of city character is substantially associated with deep emotional attachment between individuals and the city, especially peacefulness and enjoyment [53].

Sometimes, local residents feel the loss of their authentic and significant environment's character due to the changes and developments in their cities [54]. Therefore, it is necessary to preserve the valued familiar environmental features because of their importance in preserving the psychological and personal aspects as they contribute to the community perception of the authentic, significant, local and general city character [52,55].

The elements of the natural environment participate in shaping place character. The hydrological, ecological, geological, vegetation and landform features are significant features in creating the distinctive visual atributes of the place [54]. Places that have successfully integrated the natural features of water bodies, vegetation and topography maintain the natural character of the place and gain distinctive visual links [54,56].

Altman and Zube [57] show that the transformation of the undefined space to what we call a "place" comes from the transformation of the mere geographical environment by people who use it, adjust it, or give it a symbolic value, while the word 'character' indicates the set of qualities and features that identify one thing or person from others. Place character can be illustrated by aggregation of the special place features, which give their place the unique identity [54]. It is the combination of the distinctive elements of the place that give it a unique value and a special identity. These elements and features make the sense and the ambience of the place [58,59]. All the places have distinct characters, such as people who have their unique personalities that vary from one person to another [60]. However, the place character can be expressed through the unique aggregation of the socio-physical features that distinguish the environments from one to another [34,60–62]. Place character can be simply depicted as the 'sense' and the 'atmosphere' of a place [54].

The perception of town character by the public was examined by Green who focused on community perception to explore the notion of town character based on the community perspective of the small coastal town of Byron Bay, New South Wales, Australia [52–54]. Green found that there is a relationship between environmental features, their associated meanings and the perception of the character of the town. The positive meanings were distinctiveness, pleasantness, charm, familiarity, friendliness, openness, liveliness and safety. The negative meanings were boredom, ugliness, lack of charm, lack of stimulation, monotony, unpleasantness and being ordinary [52].

In his book, *The Image of the City*, Kevin Lynch (1960) reached a conclusion that places are experienced in relation to their surroundings, the events that lead to them and past experienced memories. Citizens experience the city through the experience of the small parts, niches and events that together feed the citizen with the memories and meanings [63].

Literature summary. Despite the fact that municipalities believe that good planning will lead to better economic returns, this is not always reached by the turning of land uses to successful economic and commercial uses. Land use development policies have to improve, manage and increase the effectiveness of human activities associated with land uses, with direct awareness of the importance of green open spaces and natural environmental assets [9,12,13]. Problems may be exacerbated by the lack of awareness of the social needs and problems faced by communities and affected by decisions on the community environment [64]. Some policies focused on a scope targeting environmentally and urbanely neglected areas through making these areas safe and attractive urban areas that attract investments and population alike, such as the case of Seoul, Cheonggyecheon; Riverwalk, San Antonio and the LA River project. In most of these cases the goal was not only to enhance land use, but also to sustain a town/city character by preserving a natural asset and implementing through it some purposeful planning strategies.

The tangible and physical features in the city contribute significantly to the formation of the mental images, deepen emotional attachments between individuals and the city and facilitate the process of cognition and sensory connection [53,54]. Some studies examined the city character at the scale of cities or small towns through studying the community perception [52–65]. Other studies examined the city character at the microscale, through studying the transformation of the urban identity, where the physical characteristics of the built environment are assessed [31]. But no studies assessed the impact of land use change on the city character or the impact of demolishing natural features and green corridors on converting the city character. Previous studies analyzed changes in urban identity through monitoring urban transformations [55]. They monitored morphological changes through analyzing street patterns, urban blocks and land uses [66]. Some focused on lot-based changes and redevelopment impacts by acquiring data on building permits and land use changes [67]. Some studies focused on the public perception of change independently to show the importance of integrating the stream corridor into the urban greenway plan [68].

Land use development, together with morphological foot prints, are a dynamic form of shaping and reshaping of the city. The growth of populations is physically marked by the traces of both morphological form and shape development in relation to scale and the land use proportions and land use changes. The impact of watercourses dynamics on urban morphology and land use change are major components in shaping the image of the city. Other intangible factors are sensual in the perception of people. Both the sensual and the tangible components are continually shaped and constantly affect the character.

## 3. Methodology

This research reviewed the process of land use and city character changes induced by the policy implementation of covering the watercourse and converting it to Quraysh Street. The process of evaluation went through two main stages: before and after the coverage of the stream and the assessment of the current conditions. In each stage the land use change, urban morphology change and public perception were explored.

Land use was acquired from the municipality archives, which has a collection of old maps. This research is using Bartholomew land use classification (first level) which identifies eight different land uses; residential, commercial, industrial, public and semipublic, public parks, railroads, streets and vacant land [69]. The comparison is based on land use area changes and percentage change using cad drawing area calculations on parcels. The areas are compared excluding street areas.

The morphology was registered using aerial photos by digitizing the built areas in the different years. The comparison assessed the footprint changes in terms of proportional areas of infill and fill. The research used CAD drawings to calculate the areas of the footprints.

The public perception was induced from interviews, one-to-one on-street questionnaires and online questionnaires. The three methods provided a good base for assessing changes on the city character: (1) The demonstration of opinions regarding the interviewed elderly who witnessed watercourse character changes provided the old city character; (2) the one-to-one on-street questionnaires explored stakeholders (who spent the most time at the study area) and visitors at the location; (3) the online questionnaire focused on the city character and perception of the study area by local residents and city visitors. Overall, the three methods provided a wholesome assessment of the public perception of the study area.

The first part of the first stage, the study reviewed the land use and urban morphology of the aerial photographs for the years 1918 and 1953 (Figures 5 and 6). The land use information was obtained from Amman Municipality archives. The morphology was digitized from the aerial photographs obtained from the Royal Geographic Center for these years. In order to pursue a comprehensive understanding of the situation before the coverage of the corridor, the researchers conducted interviews (N = 30) following snowball sampling with elder people (≥65 years) who lived in the period before and after the coverage of the stream (those who witnessed changes in the 1960s to the present). The sample, which was primarily from men, is especially comprised of elderly residents and workers in the downtown, especially those located in King Talal Street and Quraysh Street, who mainly manufactured small crafts or mended furniture. The first interviewee linked the researchers to another, and the second forwarded them to the third, and so on. They were tested on their perception of six characteristics: demography; history and city character; social and cultural characteristics; environment, landscape and green spaces; land use planning; and satisfaction with transportation and pedestrian networks.

In the second part of the first stage, the study reviewed the land use and urban morphology of the aerial photographs for the years 1978 and 1992 (Figure 5C,D). The Land use information was obtained from Amman Municipality archives. The morphology was digitized from photographs acquired from the Royal Geographic Center for these years. In addition, this stage succeeded in assessing public perception of the formulated character changes based on questionnaires concerning: population growth and land use change, character change, potential rehabilitation ideas and prospect solutions for the current conditions. The study conducted one-to-one, randomly distributed on-street questionnaires (N = 200) with diverse age groups (20–75) including everyday users, workers and residents of downtown. The poll took place during the month of November 2018. In the end, 65% of the sample was collected from shop visitors on both sides of Quraysh Street and King Talal Street. The shoppers' testimony concerning infrastructure, land use change and character changes were valued in this research.

In this stage, the study analyzed land use and urban morphology based on the information and illustrative plans of these periods. Maps and plans were obtained from the Greater Amman Municipality archives and publications sometimes, and then analyzed by the authors.

The second stage reviewed the current situation. This included a comparison between the existing and planned uses, to determine the extent to which municipality policies improved land use. In order to assess land use change, researchers conducted a manual land use classification of current conditions based on field observations (2018). They compared the current land uses to the previously specified land uses by the municipality policy (1960s). There were limitations in this issue due to the inaccessible municipal data concerning building permits.

In addition, the researchers prepared an online questionnaire (N = 681, ages 18–65) to explore the current perception of city character changes among people who did not witness the location prior to the corridor coverage. Due to the previously faced difficulties with volunteers to answer the on-street questionnaire, the online was foreseen to access more people willing to take the time to answer a short survey. This was done online through posting it on Facebook. It included residents of Amman and outsiders. Residents comprised 45% of the subjects. The collected responses were in the months of February and March, 2019.

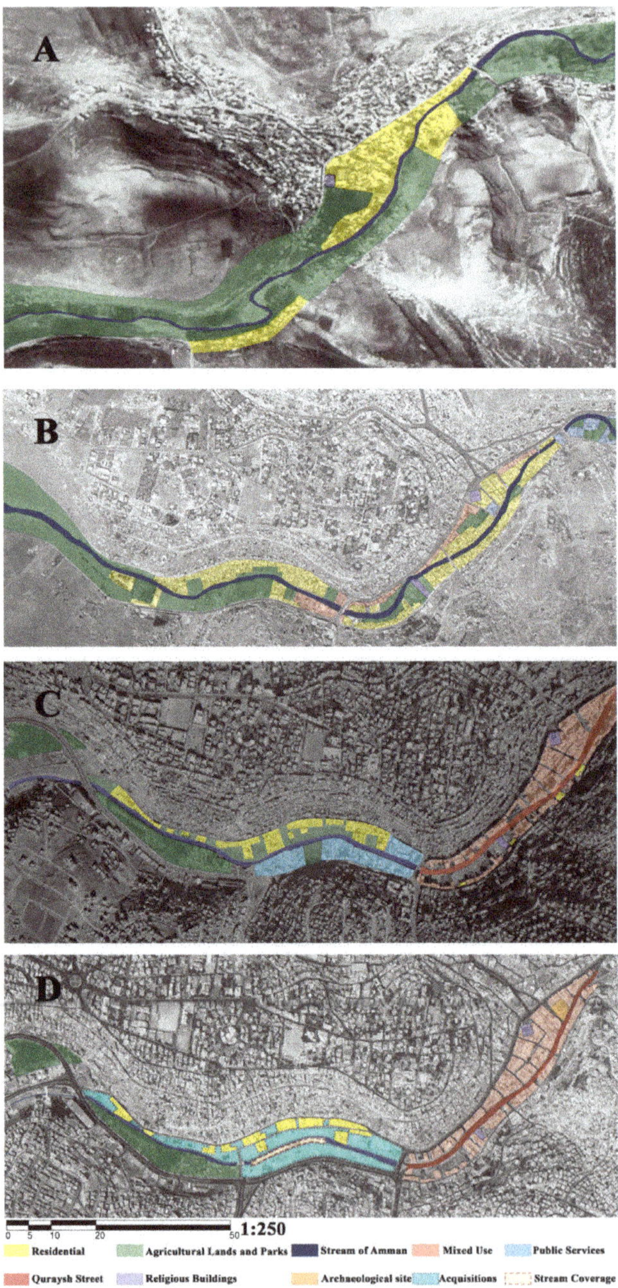


**Figure 5.** Land use, **A**: 1918, **B**: 1953, **C**: 1978, **D**: 1992 (Greater Amman Municipality archives, analyzed by researchers).

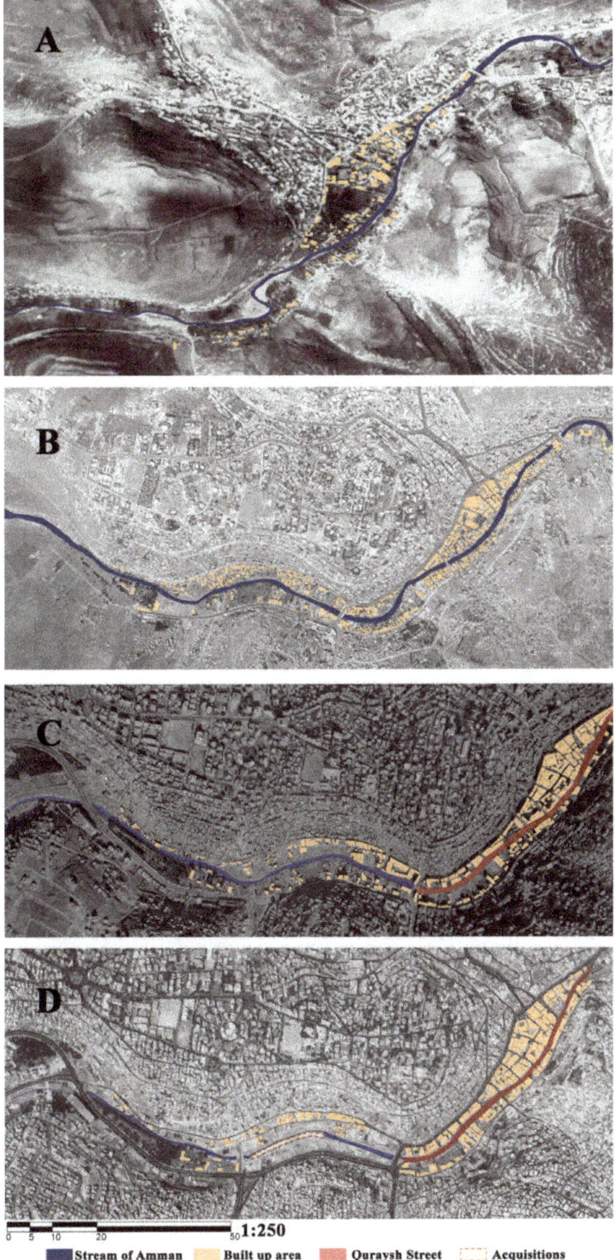



**Figure 6.** Urban morphology stages, **A**: urban morphology of 1918, **B**: urban morphology of 1953, **C**: urban morphology of 1978, **D**: urban morphology of 1992 (Aerial photographs purchased from the Royal Jordanian Geographic Center and modified by the authors).

## 4. Results

This research divided the results into two main stages: the comparison between the periods of before and after the coverage of the stream and the current conditions. In each stage, the results were divided into three topics: the land use change, urban morphology change and the public perception change.

### 4.1. Stage 1: Comparison Between the Periods of Before and After Coverage the Stream

#### 4.1.1. Land Use Changes

The results showed that there were gradual changes in land use between the years 1918 and 1953. The majority of the parcel uses in 1918 were agricultural and recreational uses (75%) and residential uses (24.3%) that was concentrated around the stream corridor (Table 1). During that period, the stream had one crossing bridge at the amphitheater site (Figure 6A).

**Table 1.** Land use change before and after the coverage of the stream.

| | Before the Coverage | | After the Coverage | |
|---|---|---|---|---|
| | 1918 | 1953 | 1978 | 1992 |
| **Land Use** | Percentage | | | |
| Agriculture/Recreational | 75.0 | 47.4 | 30.0 | 23.4 |
| Residential | 24.3 | 39.7 | 15.3 | 9.3 |
| Mixed Use | 0.0 | 8.1 | 34.8 | 35.2 |
| Public Services | 0.0 | 3.6 | 18.6 | 0.0 |
| Religious | 0.7 | 1.2 | 1.3 | 1.3 |
| Archeological sites | 0.0 | 0.0 | 0.0 | 1.2 |
| Acquisition | 0.0 | 0.0 | 0.0 | 29.6 |
| Total | 100 | 100 | 100 | 100 |

In 1953, the city developed and expanded after it became a trade destination for the population from inside and outside Jordan, especially after the Palestinian war in 1948. The map shows a slight reduction of agricultural and recreational lands to about 47.4%, and increase in the residential lands (39.7%), which resulted in creating new land uses such as mixed use (8.1%) and public services (3.6%), in addition to a slight increase in the religious buildings (Table 1). Two additional bridges occurred in that year; one within the Al-Muhajereen neighborhood and another one midway between Al-Muhajereen and the amphitheater (Figure 5B).

The urbanization overtook agricultural lands where the proportion of the agricultural and recreational lands decreased. Residential and mixed land uses spread on the edges of the main streets along the corridor side (Figure 5B).

Upon the stream coverage, a street (Quraysh Street) was erected on top of its course. Most uses on both sides of Quraysh Street converted from residential, recreational and agricultural uses to commercial and mixed land uses to inflate from 8.1% in 1953 into 34.8% by 1978. The proportion of agricultural and recreational uses decreased to 30%, and the proportion of residential use to about 15.3%. Mixed uses increased to about 34.8% (Table 1). Accordingly, the municipality added governmental services upstream consisting of 18.6% of parcel areas (Figure 6C). The view was to turn these into central business districts to support the economic development in the CBD. The watercourse is covered between the Al-Muhajereen and the amphitheater. This stretch of the stream is about 1.4 km long.

In 1967-1992 governmental acquisition started to take place upstream; houses were demolished and a part of the stream was covered and converted to a culvert. The area (land acquisition) was later transformed to the Greater Amman Municipality headquarters (Figure 6D). The agricultural and recreational uses decreased from 75% in 1918 to 23.4% in 1992. The mixed land use stayed almost the same, but the land acquisition consisted of about 29.6% (Table 1).

Table 1 shows the summary of the comparison in land use change before and after the coverage of the stream.

### 4.1.2. Urban Morphology

The results of the urban morphology analysis showed the following:

In 1918, the corridor dominated the configuration of the urban fabric. The urban blocks (fill) consisted of about 3.4% of the study area; the urban blocks and street pattern were shaped organically and intertwined with the stream corridor. There was only one main street parallel to the stream now called (King Talal Street). In addition, there was a bridge which connected the Roman Amphitheater and the agricultural fields on that side with the residential areas and mosque (Figure 6A).

In 1953, the corridor was still the most dominating element on the urban morphology, despite that the urban blocks increased (15.3%) and the number of streets and bridges increased over the corridor. The number of buildings significantly increased (about fivefold since 1918) on both sides of the corridor with the presence of urban spaces between buildings, which promoted the social activities of the residents (Figure 6B). The stream at this stage is still taking its course despite the fact that the green body of trees and agricultural fields have decreased to half. No stream sit backs were maintained to protect the course of its running stream.

In 1978, a large part of the stream (almost half) disappeared, and was replaced by a wide main street (Quraysh Street). The urban blocks consisted of about 21.3% of the study area, and these urban blocks and the streets were shaped in a loose grid pattern. The numbers of streets increased while the bridges over the stream disappeared, since the stream no longer existed for the most part. The transportation networks gave the priority in shaping the urban fabric to the vehicles. The stream coverage policy enforced land acquisition and building demolition processes, changing land use along the stream bank (Figure 6C).

The morphology of 1978 is much denser now, leaving fewer spaces between buildings (Figure 6C). The places that look vacant are those that will be obtained by land acquisition acts in the future. More very dense areas surround the watercourse at this time. This is due to the large number of immigrants coming from Palestine at this time fleeing the wars.

In 1992, Quraysh Street was the most dominant element in the urban morphology of the study area. At Ras Al-Ein, part of the remaining stream was transferred to a culvert which led to a discontinuity of the corridor (Figure 6D). In that year, the urban blocks consisted of about 25.7% of the study area, although there were demolitions of existing buildings for the purpose of governmental acquisition. Table 2 showed the summary of the comparison in urban morphology change before and after the coverage of the stream

**Table 2.** Urban morphology changes before and after the coverage of the stream.

| | Before the Coverage | | After the Coverage | |
|---|---|---|---|---|
| | **1918** | **1953** | **1978** | **1992** |
| **Fill/Infill** | Percentage (%) | | | |
| Fill | 3.4 | 15.3 | 21.3 | 25.7 |
| Infill | 96.6 | 84.7 | 78.7 | 74.3 |

### 4.1.3. Public Perception

To understand the public perception's comparison between 'before the coverage' and 'after the coverage', the researchers interviewed two groups: group A and group B. Group A resulted from a target group of interviewees of 30 persons of elder people (above 65 years old), who used to live or work in the downtown of Amman before the coverage of the stream. Group B resulted from 200 persons of visitors and residents in Amman aging 20–75 years old who witnessed Amman after the coverage of the stream. Group A, who were mainly men, used to live in the downtown, except 5% of them who used to live somewhere else, but they were working in the downtown since the 1950s. They used to work in handicrafts, agriculture, ranching, carpentry and leather tanning, and some people worked in the railway station. Although the education at that time did not exceed the middle or high school, they were learning their crafts by practice.

In describing the history and the city character features, group A described the old city of Amman as a small village, intermediated with a small river, i.e., the stream of Amman. They also described the stream banks containing orchards and some houses with a low number of population. In their opinion, the most important features of the city were the Roma Theatre, The Nymphaeum, and the Roman Arches over the stream that were removed upon covering the stream. Although they remembered the life around the stream, but they did not consider it as a main feature of the city.

On the contrary, the majority of group B considered the stream a historical value that played a significant role in the emergence of the city, and considered that the character of Amman had changed after covering the stream. They acknowledged the change of the land use and they considered it as a negative impact of covering the stream. They complain about the possibility of mixing the wastewater with rainwater. Only 41% of group B saw that covering the stream solved the transportation problem.

Concerning the social and cultural characteristics, group A expressed that social life was more intimate, simple and enjoyable then. The community consisted of groups of extended and nuclear families with strong social ties, and Amman was described "as a center of social mix and cultural gathering," where the community was a mix of Jordanians, Palestinians, Circassians, Syrians and Yemenis. In addition, they pointed out that the existence of the railroad (with traders and pilgrims) had an impact on the generated milting pot.

On the other hand, the majority of group B considered the downtown of Amman a cultural and touristic destination, and they believe that the present social life of the people in Amman is different from the past social life. Yet, they believed that there were insufficient places downtown for cultural activities, and they deemed that the stream provided an opportunity for people to meet and socialize.

When group A was asked to describe the city of Amman in the spring, they were evoked to describe the stream's nature and significance for the city; they described it then as "a green paradise" referring to the local trees surrounding the stream. The only negative image they recalled was the winter floods of the stream, and the pollution caused by discharging wastewater into the stream by residents.

Group B could not see any paradise, and most of them suffered from the lack of green spaces in the downtown. They deemed that there were no suitable places for relaxing and recreation downtown. However, they saw in the restoration of the stream a potential for a linear park that could reintroduce the nature into the downtown of Amman, and they believed that restoring it may improve the water management, reduce the runoff in the winter, reduce the air pollution and reduce the urban heat island effect.

The awareness of the land use change was recognizable in both groups A and B; group A indicated that until the seventies of the last century, Amman was an agriculture-dependent city with land orchards and fields. The shops scattered only along King Talal Street, and the residential land uses concentrated around the stream.

Group B touched the land use change and considered the rapid processes of urbanization, transportation and migration hindering land use and environmental planning processes. Most of them supported the idea of restoring the covered stream; they believed that it would provide a recreational development instead of the limited small and light industries in the area.

Both Group A and Group B agree that crossing the stream or the street is an issue; although the number of cars and streets were few before the coverage of the stream. They recognized the difficulty in accessing the other side of the stream bank; residents used to cross on wooden and metallic bridges to reach the stream banks, and to carry out their daily activities. Group B believed that covering the stream did not solve the problem, because crossing Quraysh street is still difficult.

### *4.2. Stage 2: Assessment of the Current Situation*

#### 4.2.1. Land Use Changes

To find out the improvement in the downtown environment, it was necessary to compare the planned uses with the existing uses, to know the orientation of the development plan. The Amman Municipality land use plan of the downtown was approved based on the 1967 initiative, which consisted of about 67% central business land use, 25% commercial use, 5% recreational use and about 4% of residential use (Figure 7). This plan was a wishful plan to create central business and central commercial uses in the downtown area. Its form predicated on replacing the stream corridor with a culvert topped by a traffic pine thinking this will eventually endorse the proposed uses. This plan did not leave enough room for recreation in the place most suited for recreation in the downtown area. It was also depending on the hope that once this traffic spine was created, affluent business will flourish.

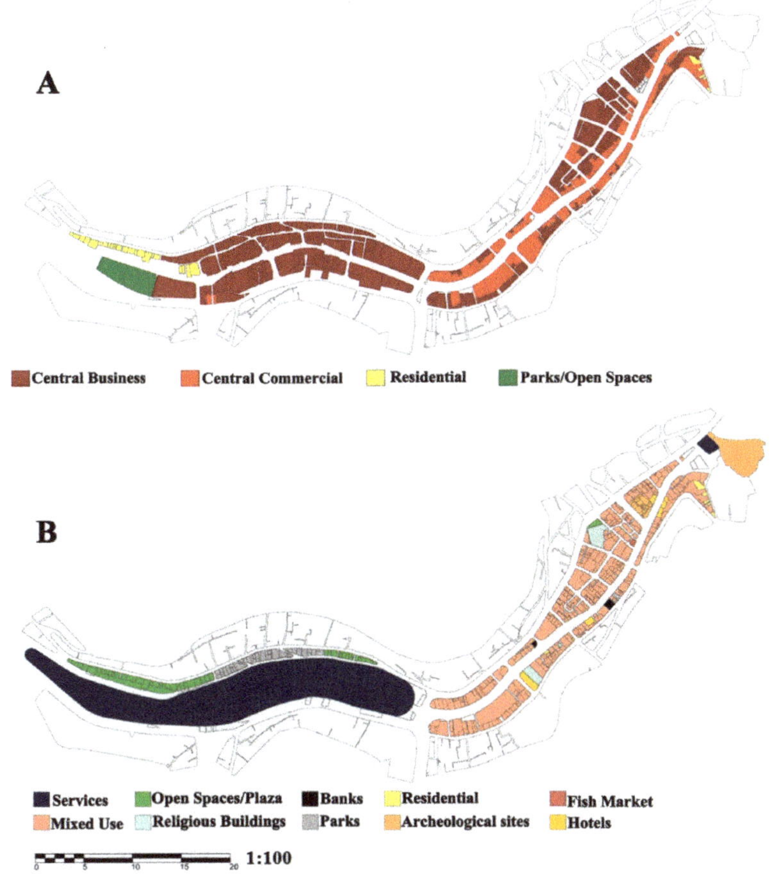


**Figure 7.** Land use plans, **A**: municipal land use plan as intended in the 1960s policy, (Source: Greater Amman Municipality; recolored by author). **B**: existing land uses (documented by the researchers).

However, to give a more detailed understanding of the current land use distribution, the classification went to a more detailed land use surveillance based on an onsite observations study. The current uses consist of about 46% mixed uses divided between light industries, markets and wholesale shops; 39% services; 5% archeological sites; 4.4% open spaces; 2.4% parking lots; 1.4% religious buildings; 1% hotels; 0.3% banks; 0.5% residential use; and 0.1% fish market (Figure 7). The services are the areas that are owned by Greater Amman Municipality (GAM). This area includes the municipality building and its plaza, Jordan Museum, Hawa Amman Radio Station and the Cultural Center. Mixed uses currently include crafts, such as furniture design and renovation; shops for mending shoes; shops for selling used clothes and shoes; workshops for making wool carpets; and some meat and candy wholesale shops. Service cars (taxis) queue along the main street to seek passengers from or to this area coming from various parts of the city. The street becomes congested and the land uses are contradicting with one another.

The aim of the improvement initiatives was to attract central commercial uses such as offices, commercial companies and institutions to make the downtown the commercial center of the capital, but the existing uses showed the polarization of mixed uses of light industrial and some of unsuitable crafts in the downtown (Figure 7).

### 4.2.2. Urban Morphology

Currently, Quraysh Street is the dominant axis on the urban morphology of the study area, where the urban blocks and streets are shaped in a loose grid pattern. The urban blocks are relatively overcrowded with an absence of recreational areas and open spaces. The urban spaces between blocks are very few, which spaces consist of streets only (Figure 8). The percentage of fill and infill now is about 37–63%, respectively.

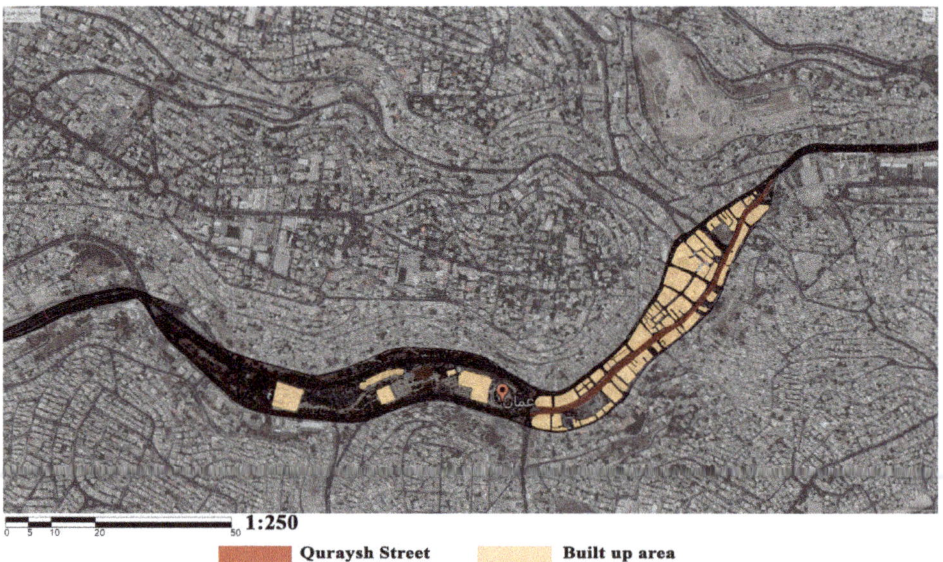


**Figure 8.** Urban morphology showing fill and infill building footprints, 2018 (Aerial photo excerpted from Google Earth 2018).

Urban areas surrounding the stream watercourse continued to become denser with fine grain agglomerations in all directions (Figure 8). The culvert is now almost finished, and very little natural green character can be noticed at the west end of the corridor. This end remained unbuilt until this day due to the lack of funding for the future King Abdullah II House of Culture and Art (Opera House), which was designed by Zaha Hadid back in 2010.

### 4.2.3. Public Perception

The detailed online questionnaire was filled by 681 participants, where 33% of the participants were males, and 67% of the participants were females. 85% of the participants were under the age of 45 years, and 15% of the participants were over the age of 45 years.

Regarding the public perception of the current character of Amman, the perception of the participants of the online questionnaire varied between "Amman is a city of traffic congestion" and "The city of Amman is rich in natural resources". The highest agreement for the current character of the city of Amman was 97.3%, where they agreed with the perception of "Amman is a city of traffic congestion". The participants agreed with at least one of the following "Amman is a city of all nationalities", "Amman is the city of the Seven Mountains", "Amman is a city of cement and stone blocks" and "Amman is a city rich in culture" by more than 80%. The current political, economic and social circumstances affected the public perception. As for the perception related to "Amman is the water city", 81.7% of participants considered that this perception is not related to the current character of the city of Amman, as the city suffers from the lack of water resources at present. The lowest percentages were for "Amman is the city for tourism", "The city of Amman is rich in recreational areas", "Amman is a city of brotherly love" and "The city of Amman is rich in natural resources".

In the open-ended online question concerning Amman's old, current and future character, the study focused upon repeated answers that received more than 30 repeats (Table 3). It shows that they acknowledged its old beauty, importance and historical character. They agreed that it does not have a current specific character, and that it is a symbol for traffic congestion. They developed concerns regarding its future character.

**Table 3.** Online, open-ended question results regarding the character of Amman; old, current and future.

| The Perception of Amman's Character | Description | Number of Repeats |
|---|---|---|
| Old | Beautiful historical character | 60 |
| | City of civilizations | 60 |
| | Multicultural character | 32 |
| Current | Lost its character | 52 |
| | Without any specific character | 72 |
| | City of traffic congestion | 34 |
| Future | Will lose its character completely | 38 |
| | Will develop an unknown character | 42 |

The perceived results confirm the physical studies and show the importance of bringing back the ecological character Amman once had.

## 5. Discussion

This study provided a review of city character change focusing on land use change, urban morphology and a public perception of Amman's covered stream. The study found the changes in land use, urban morphology and city character were induced due to policy implementation, in addition to the population escalation followed by contiguous urban growth in the downtown area.

Before covering the stream, the land uses were confined to agricultural, residential and recreational uses, with a few commercial uses and services spread along the stream banks. After the implementation of the 1960s policy, land uses had evolved on both sides of the corridor (the covered stream), increasing the size of commercial, services and mixed land use type of buildings.

The presence of the stream played a vital role in distinguishing a socioeconomic lifestyle along the stream banks where the city had a small, simple and multicultural society. Residents were working in agriculture, handicrafts, trading and fishing. Residents' houses were scattered alongside the stream, where bridges over the stream helped them to access their daily destinations. The stream of Amman was the dominant element of the urban fabric before the 1960s, showing a primarily green corridor with urban spaces surrounding the stream. The stream enhanced the recreational and social life of the residents in addition to some historical and religious buildings.

The urban morphology of the study area converted from an organic and a spontaneous scatter into a loose-grid pattern dominated along the main street corridors. The dominant element in Old Amman has been changed from a green corridor into a traffic corridor that attracted emerging land uses. The land use change also affected the urban morphology, making the urban blocks more intensive, while allowing for small crafts shops and low budget uses to coexist. The urban spaces between urban blocks decreased slowly until the stream disappeared completely. No signs of a green corridor remained to remind people of the old city life.

The culvert created a more urbanized downtown with mixed industrial land uses concentrated on both sides of the covered corridor. Some important land uses disappeared, such as agricultural and recreational uses. Unlike the 1955 master plan which proposed green fingers for the green infrastructure of the city, the 1960s policy robbed the city its green corridors. This is an important indicator of the need to make changes for the CBD in order to revise and reverse this policy.

The proposal of the 1960s policy of creating central business in the downtown was not met in the actual current land use. The impact of change in the land use inspired more traffic congestion. Consequently, the initial problems of traffic within the area were not solved. Floods remained as a problem well into the present time. While the policy of the 1960s intended to solve traffic congestions and winter floods for a more prosperous downtown, the results have developed otherwise.

The land use change affected the social and perceptual aspects of the city character. The study found a dramatic change of community perception between the coverage stages. The Elderly who lived in the city before the coverage of the stream and witnessed all stages of change presented the perception of the city before the coverage of the stream. Their perception of the city reflected a small city flourished with history, diversity and civilization. The city was rich in natural resources, such as the stream and abundant springs. In addition, the city was rich in biodiversity, including trees, wildlife, fish and livestock. The downtown or the city center was the political, economic, social and cultural center, which was frequented by people from all regions.

The perception of the city after covering the stream was presented by the point of view of those who answered the on-street, one-to-one questionnaire and the online questionnaire. Although the participants considered the downtown as a tourist and cultural destination, most of them had no favorite destinations in the downtown. The participants of the survey focused more on the negative effects in the downtown area, such as the absence of green areas, the increased visual pollution and runoff rates.

The online questionnaire revealed the perception of the present city character as "the city of traffic congestion" (97.3%). This is a relatively large percentage, and is an alarming negative indicator of the overall perception of the city character. The perception of "Amman is the city for all nationalities" reached 87%, which is an added value. Another negative perception with high percentage was "Amman is the city of concrete and stone blocks", which was 85%. However, 82% of the participants considered that Amman was NOT "the city of waters", which is a negative, alarming indicator, as a result of the policy implementation and the lack of knowledge concerning the history of the city. These findings regarding the general perception of the city character raised concerns about the future of city character. The present conditions, they believed, showed character loss or absence. Some of the participants indicated that the future city character will be lost or will develop an unknown character.

Community's perception of the city character varies depending on the age group. The perception of the generations who lived through the changes differs from the generations who lived after those changes. The generations who had experienced the change since its earliest stages conceived the place in relation to their personal experience, the physical elements (natural and environmental ones), the social and economic aspects. Generations who did not live through the change have viewed the character based on an economic, urban and architectural perspective not in relation to the natural environment and the context of the city kernel and surrounding region. They acknowledged its historical and multicultural character, and however thought that its character is now lost and replaced by a harsh traffic and building-dependent city.

All the results related to the land use change and to the perception change of city character indicated that we need to reconsider the adopted policies. In addition, we need to know how to implement them, taking into consideration all environmental, physical, social and economic factors to restore and preserve the city character. Additionally, the future development of the city should be commensurate with the needs of the current and future generations in a sustainable manner. This is especially important since many people had concerns for losing the chance to save the significant character of Amman. Interfering with ecological or environmental urban issues may create bigger problems that are hard to control, given the circumstances.

In the case studies of Seoul and Los Angeles, the solutions for the mistaken policies treated the conditions as an opportunity to create positive changes. Similarly, the conditions in Amman can benefit from such opportunities for creating positive changes.

Although the study focused on physical and perceptual factors as affected by a certain engineered policy to mitigate flood and traffic problems, its results can go beyond these boundaries to understand the way a policy is formulated and implemented. It can also help in understanding that municipal power, environmental assets, sociocultural needs and city character are interrelated components of change. There are many lessons that can be learned from this exploration of policy implementation and change. The results show many hidden values that can be deduced from this case study:

Land use development towards central business and vital commerce cannot be motivated by policies that eradicate an environmental asset that once contributed to the urban character. Sometimes, as in this case, an engineered problem solving is not always the solution for a more prosperous city center. This was also true in the case of the LA River.

Urban morphology can eradicate urban spaces once policies neglect the proper setback and allocation of open spaces and lose respect for environmental assets. With the absence of strict policies, densification can get out of hand.

Change happens incrementally over so many years, that the perception of change can sometimes pass unnoticeable. The new generations sometimes fail to see the actual original sense of a place and what it once presented due to urban policies that hassle to create physical changes without taking into account the public needs or the historical and heritage value of a place.

Young and old generations are accepting of the importance of the natural environment in this context for the lack of open spaces and recreational environments. They like to see changes that bring back more of nature and less of traffic congestion. They are enthusiastic about environmental transformations. This is true despite the fact that the new generation was not aware of the stream corridor and did not know that it had a substantial influence on the historical city character as the "City of Waters".

Flood management and traffic control policies ought to consider the public perception of the place and ecological factors as other determinants of change. Sometimes with some creativity, problem-solving can create a new welcoming nature that has the potential to transform—in a parallel way—the economy and nature such as the examples of San Antonio Texas and Cheonggyecheon in Seoul. In these two examples, the problem-solving brought a new character to the city that can grow, change, and elevate its economic and infrastructure planning.

The power to create improved city centers is mostly in the hands of municipal management and action plans. Therefore, the awareness of what physical changes can do to the city character should be enhanced for decision-makers at the municipality level.

The proposal of bringing back the stream into an urban greenway was accepted by the local community positively. The study revealed the desire of the local community to restore the stream of Amman in an urban greenway. They justified the several factors, including the need to create new development prospects in the downtown, the need to restore the unique historical character of the city, the need for recreational areas and green spaces suitable for all groups of community, the need for suitable places for cultural activities and the need to conserve environmental resources. They defended the need to mitigate the environmental impacts of the urbanization process such as reducing the runoff, helping in storm water management, reducing the impact of urban heat island and reducing the air pollution. Bringing back the stream into an urban greenway is an opportunity which will provide a linear green heart that will restore the character of the stream. It will also provide an attractive and safe environment for pedestrian movement. It has the potential for changing the lifestyle in the city from car-dependent to walkable streets and green networks. This, in turn, will encourage investments in the downtown, which will subsequently provide new land uses and new job opportunities more fitting with the capital city CBD of Amman.

## 6. Conclusions

The purpose of this research was to understand changes in the city character as imposed by the policy implementation, in addition to exploring public opinions regarding that change. Accordingly, the research performed a longitudinal study that focused on the history of the city and its flowing stream as a kernel for urban settlement and the procession of urban planning policies and master plans focused on the stream and the downtown of the capital city of Amman. Then it analyzed layers of change to the land use, the urban morphology and the public perception before and after the stream coverage.

What distinguishes this research is that it has combined between Green's methodology (1999, 2000, and 2010) in exploring the city character based on community perception, and the methodology of Beyhan and Çelebi [70] who analyzed the relationship between the change of urban identity and the urban transformation through analyzing the morphological characteristics in urban built environments. In addition, it studied the city character change induced by the implementation of land use development policies, through physical, perceptual and functional assessment of the city character transformations.

This research could be considered as foundation research to assess the relationship between policy implementation and city character. This assessment methodology can be applied to the micro- and macroscales of the city to provide an understanding of the change and the factors driving it.

For the case of the Amman stream in particular, this research opens the gate to stimulate different assessment and evaluation methods to explore urban planners' points of view to fix the situation of the covered stream (Quraysh Street), or even to persuade the decision-makers to reevaluate opening up the stream. This research may open up an opportunity for further research to explore and trace the societal changes demonstrated in the small details of the lifestyle in downtown and community behavior towards the city center.

Future applicable policies need to be reevaluated to determine their success and social acceptance, and to avoid irreparable losses to the environment and the culture of the place. Physical and environmental elements of cities are important in shaping the public perception of the city character. Therefore, it is important to preserve these distinctive elements, to maintain their meanings and the memory of the place for future generations.

Both the perceptual views and the physical changes induced by the coverage support retaining city character by creating a framework for a stream corridor. Many other green corridors in the city can become part of this greenway or green corridor network in the future plans to improve urban ventilation. This research recommends a future study that can focus on the cost–benefit values for implementing green fingers strategy, as once proposed by the 1955 plan, in the city of Amman. This is important because it may have a major impact on climate change and the economic factors, which are not detailed in this study.

**Supplementary Materials:** The following are available online at http://www.mdpi.com/2073-445X/8/12/195/s1.

**Author Contributions:** Conceptualization, A.A.G. and E.M.A.; data curation, E.M.A.; investigation, A.A.G. and E.M.A.; methodology, A.A.G. and E.M.A.; writing—original draft, A.A.G. and E.M.A.; writing—review and editing, A.A.G., E.M.A., and L.B.A.

**Funding:** No funding was received for this research. However, supporting documents were provided by Amman Municipality.

**Acknowledgments:** We thank Greater Amman Municipality for providing us with photos and plans from their archives. Colleagues who helped developing and strengthen our research including Atef Nusseir and Yasmein Okour. This work was done at Jordan University of Science and Technology as part of the fulfillments of Master's thesis submission requirements.

**Conflicts of Interest:** The authors declare no conflict of interest.

## References

1. Silva, J.; Saraiva, M.; Ramos, I.; Bernardo, F. Improving Visual Attractiveness to Enhance City–River Integration—A Methodological Approach for Ongoing Evaluation. *Plan. Pract. Res.* **2013**, *28*, 163–185. [CrossRef]
2. Champion, T. Urbanization, Suburbanization, Counterurbanization and Reurbanization. In *Handbook of Urban Studies*; Paddison, R., Ed.; Thousand Oaks: London, UK; SAGE Publications: New Delhi, India, 2001; pp. 143–161.
3. Paul, M.J.; Meyer, J.L. Streams in the Urban Landscape. *Annu. Rev. Ecol. Syst.* **2001**, *32*, 333–365. [CrossRef]
4. Oke, T. The energetic basis of urban heat island. *Q. J. R. Meteorol. Soc.* **1982**, *108*, 1–24. [CrossRef]
5. Kalnay, E.; Cai, M. Impact of urbanization and land-use change on climate. *Nature* **2003**, *423*, 528–531. [CrossRef] [PubMed]
6. Wild, T.C.; Bernet, J.F.; Westling, E.L.; Lerner, D.N. Deculverting: Reviewing the evidence on the 'daylighting' and restoration of culverted rivers. *Water Environ. J.* **2011**, *25*, 412–421. [CrossRef]
7. Samuel 2, King James Bible, David Captures Rabbah, in Bible Hub. Available online: https://biblehub.com/kjv/2_samuel/12.htm (accessed on 13 December 2019).
8. Erskine, S. *Trans-Jordan: Some Impressions*; Ernest Benn Limited: London, UK, 1924.
9. Abu-Dayyeh, N. Persisting vision: Plans for a modern Arab capital, Amman, 1955–2002. *Plan. Perspect.* **2004**, *19*, 79–110. [CrossRef]
10. Al-Asad, M. Ever-Growing Amman. Urban Crossroads No. 42. Center for the Study of the Built Environment (CSBE), 16 June 2005. Available online: http://www.csbe.org/ever-growing-amman (accessed on 13 December 2019).
11. Al-Azra'e, M. The stream of Amman: From an infected swamp, to a modern construction facility. *Alrai Newspaper*, 17 September 1971; no. 93. 6.
12. Al-Kurdi, M.A.A.-S. *Amman; History, Culture, and Heritage (The Conservative City)*; Dar Amar: Amman, Jordan, 1999.
13. Al-Mousa, S. *Amman*; Greater Amman Municipality: Amman, Jordan, 2003.
14. Bank, W. Jordan Population 2019, World Population Review. 2019. Available online: www.worldpopulationreview.com (accessed on 8 November 2019).
15. Al-Allaff, R.A.; Jrew, B.; Abojaradeh, M.; Msallam, M. Evaluation and Improvement of Traffic Flow and Traffic Network Management System in Al-Shmesani–Amman, Jordan. In Proceedings of the 7th Traffic Safety Conference, Amman, Jordan, 12–13 May 2015.
16. Kloub, M. *The Jordan Times*. 28 July 2016. Available online: http://jordantimes.com/news/local/not-easy-job-amman%E2%80%99s-traffic-headache-curable-%E2%80%94-experts (accessed on 13 August 2019).
17. Roya News. 2019. Available online: https://en.royanews.tv/news/16807/2019-03-03 (accessed on 24 March 2019).
18. Roya News. 2019. Available online: https://en.royanews.tv/news/16768/Main_street_in_Downtown_Amman_ (accessed on 25 March 2019).
19. Msallam, M.; Abojaradeh, M.; Jrew, B.; Al-Allaff, R.A. Evaluation of traffic flow and traffic network management. *Int. J. Civ. Eng. Constr. Estate Manag.* **2016**, *4*, 1–21.
20. History of Jordan. 2019. Available online: http://www.historyofjordan.com/jordan2/jh/collection1.php?id=26&page=713 (accessed on 24 March 2019).
21. Malkawi, H.S. *City of Amman 1921–1947: Historical Study*; Dar Al-Kindi for Publishing and Printing: Irbid, Jordan, 2002.
22. Kadhim, M.; Rajjal, Y. City profile Amman. *Cities* **1988**, *5*, 318–325. [CrossRef]

23.  Potter, R.B.; Darmame, K.; Barham, N.; Nortcliff, S. "Ever-growing Amman", Jordan: Urban expansion, social polarisation and contemporary urban planning issues. *Habitat Int.* **2009**, *33*, 81–92. [CrossRef]

24.  Correll, M.R.; Lillydahl, J.H.; Singell, L.D. The Effects of Greenbelts on Residential Property Values: Some Findings on the Political Economy of Open Space. *Land Econ.* **1978**, *54*, 207–217. [CrossRef]

25.  Walmsley, A. Greenways and the making of urban form. In *Landscape and Urban Planning*; Elsevier: Amsterdam, The Netherlands, 1995; Volume 33, pp. 81–127.

26.  Imam, K.Z.E.A. Role of urban greenway systems in planning residential communities: A case study from Egypt. *Landsc. Urban Plan.* **2006**, *76*, 192–209. [CrossRef]

27.  Kang, C.D.; Cervero, R. From Elevated Freeway to Urban Greenway: Land Value Impacts of the CGC Project in Seoul, Korea. *Urban Stud.* **2009**, *46*, 2771–2794. [CrossRef]

28.  Jang, M.; Kang, C.-D. Urban greenway and compact land use development: A multilevel assessment in Seoul, South Korea. *Landsc. Urban Plan.* **2015**, *143*, 160–172. [CrossRef]

29.  Los Angeles Department of Water and Power, Internet Archive-WayBack Machine. 1 February 2009. Available online: https://web.archive.org/web/20090201065413/http://wsoweb.ladwp.com/Aqueduct/historyoflaa/index.htm (accessed on 20 September 2019).

30.  Linton, J. *Down by the Los Angeles River: Friends of the Los Angeles Rivers Official Guide, illustrated Ed.*; Wilderness Press: Berkeley, CA, USA, 2005.

31.  Smith, D. Angelenos' vision of their river is created from a made-up memory. *Los Angeles Times*, 15 August 2013.

32.  Guinn, J.M. *A History of California and An Extended History of Its Southern Coast Counties, also Containing Biographies of Well-Known Citizens of the Past and Present*; Historic Record Company: Los Angeles, CA, USA, 1907; Vol. 1.

33.  Carruth, A. "Kcet,". 20 March 2014. Available online: https://www.kcet.org/shows/artbound/a-brief-history-of-public-art-and-the-la-river (accessed on 21 September 2019).

34.  Hawthorne, C. Column: Frank Gehry's controversial L.A. River plan gets cautious, low-key rollout. *Los Angeles Times*, 18 June 2016.

35.  Kamal, A.; Doganer, S.; Ruvuna, J.; Flores, J.; Hernandez, E.; Nishimoto, T. Wayfinding and Accessibility in the San Antonio RiverWalk: A Model for Urban Design Education. *Int. J. Arch. Res.* **2010**, *4*, 321–406.

36.  Bauml, S. *The River Walk—From Vision to Reality*; University of the Incarnate Word: San Antonio, TX, USA, 2019.

37.  Zunker, V.G. *A Dream Come True Robert Hugman and San Antonio's River Walk*, 1st ed.; Amazon: Seattle, WA, USA, 1983.

38.  Fisher, L.F. *River Walk: The Epic Story of San Antonio's River*; Maverick Publishing Company: San Antonio, TX, USA, 2006.

39.  Fisher, L.F. *Crown Jewel of Texas, the Story of San Antonio's River SIGNED, illustrated Ed.*; Maverick Publishing Company: San Antonio, TX, USA, 1997.

40.  Cengİz, B.; Smardon, R.C.; Memlük, Y. Assessment of river landscapes in terms of preservation and usage balance: A case study of the Bartin River Floodplain Corridor (Western Black Sea Region, Turkey). *Fresenius Environ. Bull.* **2011**, *20*, 1673–1684.

41.  Institute for Transportation and Development Policy. *The Life and Death of Urban Highways*; EMBARQ. ITDP: Washington, DC, USA; New York, NY, USA, 2012.

42.  Findlay, S.J.; Taylor, M.P. Why rehabilitate urban river systems? *Area* **2006**, *38*, 312–325. [CrossRef]

43.  Palmer, M.A.; Hondula, K.L.; Koch, B.J. Ecological Restoration of Streams and Rivers: Shifting Strategies and Shifting Goals. *Annu. Rev. Ecol. Evol. Syst.* **2014**, *45*, 247–269. [CrossRef]

44.  Prior, J. Urban river design and aesthetics: A river restoration case study from the UK. *J. Urban Des.* **2016**, *21*, 512–529. [CrossRef]

45.  Fabos, J.G.; Ahern, J. *Greenways: The Beginning of an International Movement*, 1st ed.; Elsevier: Amsterdam, The Netherlands; New York, NY, USA, 1995.

46.  Searns, R.M. The evolution of greenways as an adaptive urban landscape form. *Landsc. Urban Plan.* **1995**, *33*, 65–80. [CrossRef]

47.  Miller, W.; Collins, M.G.; Steiner, F.R.; Cook, E. An approach for greenway suitability analysis. *Landsc. Urban Plan.* **1998**, *42*, 91–105. [CrossRef]

48. Shafer, C.S.; Scott, D.; Mixon, J. A Greenway Classification System: Defining the Function and Character of Greenways in Urban Areas. *J. Park Recreat. Adm.* **2000**, *18*, 88–106.

49. Gobster, P.H.; Westphal, L.M. The human dimensions of urban greenways: Planning for recreation and related experiences. *Landsc. Urban Plan.* **2004**, *68*, 147–165. [CrossRef]

50. Carmon, N.; Shamir, U. Water-sensitive planning: Integrating water considerations into urban and regional planning. *Water Environ. J.* **2010**, *24*, 181–191. [CrossRef]

51. Hoşgör, Z.; Yigiter, R. Greenway Planning Context in Istanbul-Haliç: A Compulsory Intervention into the Historical Green. *Landsc. Res.* **2011**, *36*, 341–361. [CrossRef]

52. Green, R. Meaning and form in community perception of town character. *J. Environ. Psychol.* **1999**, *19*, 311–329. [CrossRef]

53. Green, R. Notions of town character. *Aust. Plan.* **2000**, *37*, 76–86. [CrossRef]

54. Green, R.J. *Coastal Towns in Transition: Local Perceptions of Landscape Change*; Springer: Dordrecht, The Netherlands, 2010.

55. Schuster, M.D. Growth and loss of regional character. *Places* **1990**, *6*, 78–87.

56. Saarinen, T.F.; Cooke, R.U. Public Perception of Environmental Quality in Tucson, Arizona. *J. Ariz. Acad. Sci.* **1971**, *6*, 260–274. [CrossRef]

57. Altman, I.; Zube, E.H. (Eds.) *Introduction, in Public Places and Spaces*; Plenum Press: New York, NY, USA, 1989; pp. 1–5.

58. Jiven, G.; Larkham, P.J. Sense of Place, Authenticity and Character: A Commentary. *J. Urban Des.* **2003**, *8*, 67–81. [CrossRef]

59. Sepe, M.; Pitt, M. The characters of place in urban design. *Urban Des. Int.* **2014**, *19*, 215–227. [CrossRef]

60. Kasperson, J.X.; Kasperson, R.E.; Turner, B. *Regions at Risks: Comparisonsof Threatened Environments*; UN University Press: Tokyo, Japan; New York, NY, USA, 1995.

61. Verheye, W.H. Land Cover, Land Use and the Global Change. In *Encyclopedia of Land Use, Land Cover and Soil Sciences-Land Cover, Land Use and the Global Change*; Verheye, W.H., Ed.; Encyclopedia of Life Support Systems (EOLSS)/UNESCO: Oxford, UK, 2009; pp. 45–80.

62. Turner, B., II; Skole, D.; Sanderson, S.; Fischer, G.; Fresco, L.; Leemans, R. *Land-Use and Land-Cover Change—Science/Research Plan*; IGBP Report No. 35, HDP Report No. 7.; IGBP and HDP: Stockholm, Sweden; Geneva, Switzerland, 1995.

63. Lynch, K. *The Image of the City*; The M.I.T. Press: Cambridge, MA, USA; London, UK; Massachusetts Institute of Technology: Cambridge, MA, USA, 1960.

64. Albrechts, L. Reconstructing Decision-Making: Planning Versus Politics. *Plan. Theory* **2003**, *2*, 249–268. [CrossRef]

65. Gharaibeh, A.A.; Jaradat, R.A.; Okour, Y.F.; Al-Rawabdeh, A.M. Landscape Perception and Landscape Change for the City of Irbid, Jordan. *J. Arch. Plan.* **2017**, *29*, 89–104.

66. Lim, H.; Kim, J.; Potter, C.; Bae, W. Urban regeneration and gentrification: Land use impacts of the Cheonggye Stream Restoration Project on the Seoul's central business district. *Habitat Int.* **2013**, *39*, 192–200. [CrossRef]

67. Bürgi, M.; Russell, E.W. Integrative methods to study landscape changes. *Land Use Policy* **2001**, *18*, 9–16. [CrossRef]

68. Asakawa, S.; Yoshida, K.; Yabe, K. Perceptions of urban stream corridors within the greenway system of Sapporo, Japan. *Landsc. Urban Plan.* **2004**, *68*, 167–182. [CrossRef]

69. Wang, X.; Rainer, H. *Research Methods in Urban and Regional Planning*; Springer Science and Business Media: Berlin, Germany, 2008.

70. Beyhan, Ş.G.; Gürkan, Ü.Ç. Analyzing The Relationship Between Urban Identity And Urban Transformation Implementations In Historical Process: The Case of Isparta. *Int. J. Arch. Res.* **2015**, *9*, 158–180. [CrossRef]

*Article*

# The Role of Vegetation in the Morphological Decoding of Lisbon (Portugal)

**Rui Justo [1,*] and Maria Matos Silva [2]**

[1] Formaurbis LAB, CIAUD—Research Centre of Architecture Urbanism and Design, Lisbon School of Architecture, Universidade de Lisboa, Lisboa 1349-063, Portugal

[2] URBinLAB, CIAUD—Research Centre of Architecture Urbanism and Design, Lisbon School of Architecture, Universidade de Lisboa, Lisboa 1349-063, Portugal; m.matosilva@fa.ulisboa.pt

* Correspondence: ruijusto@campus.ul.pt

Received: 20 December 2019; Accepted: 9 January 2020; Published: 11 January 2020

**Abstract:** In the academic context, especially in the fields of architecture, landscape architecture, and urbanism, urban form studies are assumed to be a vehicle for reflection on the built and unbuilt city. This essay aims to challenge the most common and stabilized morphological approaches in the city reading process, invoking vegetation and its role as an element of urban composition that is recurrently left out of it. Methodologically, this work uses the city of Lisbon to carry out a morphological characterization of different homogeneous areas based on a decomposition process of urban systems and elements. The article focuses on the reading of the public component of three homogeneous areas in Lisbon—Alfama, Avenidas and Alvalade—and specifically on the role of urban greenery as a systemic element of the formal or informal composition of the city. Through an initial systematization process reflects upon the formal attributes of vegetation and trees in particular, this research may contribute not only to the development of the discipline of urban morphology applied to the city of Lisbon but also to the acknowledgment of urban greenery as a contributor to the creation of specific, unique, and unrepeatable spaces within urban landscapes.

**Keywords:** urban morphology; vegetation; Lisbon

---

## 1. Introduction

This article aims to question the role of vegetation in the understanding and design process of the city through the morphological characterization of different homogeneous areas extracted from the urban fabric of the city of Lisbon. The approach to each urban segment is based on a morphological decomposition process of urban systems and elements. This enables the understanding of the complexity of each urban unit from the reading of its public and private components. The reading of the private city is made from the interpretation of the built fabric and the elements that structure it, such as urban blocks, plots, buildings and courtyards. The public city, as addressed in this article, is understood from the reading of the urban layout as a bi-dimensional representation of the city's public spaces and elements, such as streets, squares and vegetation as a variant and leading element of this exercise.

By taking vegetation as an element of urban morphology, this work challenges the most common morphological approaches to the city that invariably relies on stabilized and tested methodologies based on different schools of urban morphology such as the Italian, Anglo-Saxon, and French ones.

## 2. Urban Morphology: From City to Urban Greenery

Since the mid-twentieth century, city forms have taken a central role in the understanding of the dynamics and complexity of urban fabric.

Camillo Sitte [1] and Joseph Stübben [2] were the first urbanists to contribute towards the systematic study of the city form, later known as the discipline of urban morphology. Between the Great Wars, geographers from the German and French schools, and the Italian architecture school, also began to work on this discipline, with a particular following in France [3].

These practitioners (namely, Muratori, Caniggia, and Maffei) call themselves typologists as they reveal the urban physical and spatial structure based on a detailed classification by the types of elements that compose it: "they study the pieces or cells—buildings and open spaces contained within the framework of a discrete piece of land in single ownership or use—that generate and change the cityscape" [4].

The typo-morphological analysis aims to individualize each element, recognizing and systematizing its characteristics, its differences, and its relationship with the urban context, historical period, and the society that originated it [5]. Yet, as argued by Aymonimo [6], the identification of the type and its corresponding typology should not be solely understood as a methodical act of classification, but also as an important design tool. Not only does it decode the city's physical and spatial structure, but it also systematizes the inherent processes of consolidation and design [7]. It is, therefore, an important tool to consolidate ideas and concepts.

Lynch [8] and Cullen [9] also read and perceived the city and its built environment from the decomposition of its morphological elements, such as streets, squares, buildings and topography, amongst others [8,9]. They compared, connected, and related each morphological element of urban fabric in each context, and with each other, to contribute to the explanation of the city. Through this method of urban fabric decomposition, the relationship between the different elements that form the city through history is more easily interpreted [10].

The morphological analysis is a classic reading tool of the city that simplifies what is naturally complex. These procedures, in their different approaches, allow the understanding of the whole through the reading of each element, their structural relations and interaction over time.

When considering the specific identification and characterization of the different types of vegetation forms within the city, two references must be highlighted: "L'arboriculture urbaine" [11], a work by Laurent Mailliet and Corinne Bourgery, and "L'urbanisme vegetal" [12], authored by Caroline Stefulesco. Both references contribute to initial identification of the most important principles of the composition of urban greenery. In fact, the former highlights "the many ways to compose vegetation and combine its effects" [11] (p. 67):

(1) Regular or Random: Regular plantations usually emphasize rectilinear tracings and other geometric figures inspired by classical architectonic compositions. Random plantations introduce an irregularity into the city that is intended to be natural, yet they require significant know-how.

(2) Volume: Vegetation, even leafless, constitutes volumes comparable to architectural structures. Vegetation monuments allow tree compositions of exceptional dimensions.

(3) Dome: Foliage may constitute domes that define interior spaces, sometimes with strong architectonical features. They form a converting that tempers the excesses of heat or luminosity.

(4) Border: When vegetation borders set the boundaries of space. They consolidate and explicit urban textures. In the proximity scale, they create privileged places.

(5) Staging: When vegetation arrangements emphasize and value sights and landscapes as well as buildings and monuments, framing them on a regular or irregular basis.

(6) Apparatus: When vegetation amplifies the staging by overflowing the composition beyond the venue, building, or monument;

(7) Accompaniment: When vegetation planted in the domains contiguous to the public space participates in an important way in the collective landscape. Often, vegetation overflow completely covers the walkways.

(8) Landmark: When a tree or vegetation complex is notable for its size, architecture, flowering or foliage, and contributes to orientation and location.

(9) Covering: When climbing or suspended vegetation covers buildings, especially when the necessary conditions for its growth have been integrated into the architectural proposal [11].

This systematization is relevant as a principle for the understanding of the multiple roles and forms that vegetation can have in the construction of urban space. While this previous reference combines the formal vegetation aspects of element appearance with other aspects of specific positioning and urban space definition, the work developed by Sérgio Proença [13] on the streets of the city of Lisbon offers a more simplified view, solely focused on trees. By separately addressing the different aspects that determine the relationship between tree planting and street space—the conformation of its layout, cross-section, and partition—and the associated typological synthesis, Proença identified four types of aggregation and distribution of urban greenery:

(1) Unique elements: Trees that are placed sporadically and in an isolated way in the city, reinforcing the uniqueness of a given context.

(2) Simple alignments: The most common form of street layout partitioning, by distributing trees of the same type in single or multiplied lines and a lateral or central position to the street space.

(3) Complex alignments: Alignments defined by different tree species.

(4) Sets of assemblages: The spaces of a more informal nature, composed by the aggregation of trees that are arranged organically, forming a green mass.

The analysis of both references serves the work that is presented here, not only in the understanding of the formal potentialities of vegetation forms within the city but also in the specificity of trees as an element of street spatial composition and in its systematization in particular types that explain the phenomenon.

Bearing in mind that Lisbon (Portugal) is the chosen case study, two additional references are explored: "The Tree in Portugal" [14] and "Trees in the City" [15]. These works offer better knowledge about the types of vegetation formations in the Portuguese context and their circumstances in Lisbon's urban system.

## 3. Lisbon's Urban Form: Alfama, Avenidas, and Alvalade

In Lisbon (Figure 1), the heterogeneous nature of urban space is the result of formal characteristics that are representative of the intersection of different eras and peoples, contextualized by its privileged situation in the territory. Lisbon has always profited from its strategic positioning, essentially given by the navigability and natural harbour conditions of the Tagus Estuary, which promoted human presence and the establishment of open ports for all. The Atlantic history of the city and the landscape from which it emerged, combining the Tagus Estuary with an irregular topography dominated by a set of hills and valleys, determined the medieval occupation of Lisbon. This occupation developed from a dual need to accommodate the urban fabric to the topography whilst benefiting from a safe and flourishing port. Its history is marked by a succession of earthquakes such as the one in 1755 that forever transformed its image, creating a new identity, defining new boundaries, and further propelling the emergence of new neighbourhoods. Like other nineteenth-century European cities, avenues ripped open, expanding the city northward and along the valleys. The city's growth would later cover all directions, driven by the twentieth-century demographic boom, which justified the emergence of new developments.

In Lisbon, the tension between diversity and unity is evident due the existence of urban fabrics that has undergone long processes of sedimentation (Alfama), urban fabrics that result mainly from an idea of urban production based on the conception of public space (Avenidas), others designed as an integral unit (Alvalade), operations which reject the classic elements of urban composition such as the street and block (Olivais), or more recent urban fabrics representing the recovery of these elements in city urban composition (Parque das Nações). All of these emerged between different development phases of the city and among different natural land morphologies such as hills, valleys and the riverfront (Figure 2). The combination of urban fabrics in the city not only translates into its formal complexity, but also into the dialogue between its public form and the presence of trees (Figure 3).

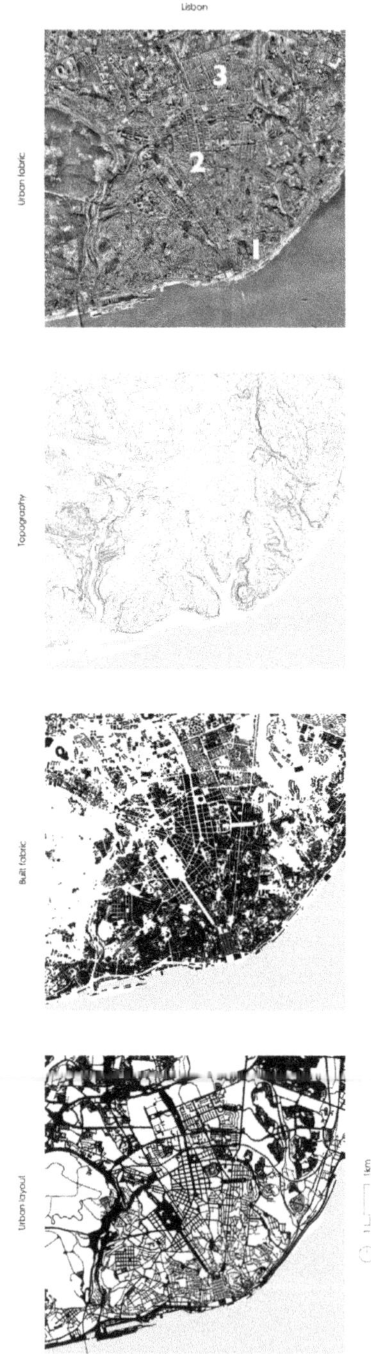



**Figure 1.** Lisbon's urban morphology. Identification of the three homogeneous areas analyzed: (1) Alfama, (2) Avenidas and (3) Alvalade.

**Figure 2.** Morphological table of the three homogeneous areas analyzed: Alfama, Avenidas and Alvalade. Photographic source: Filipe Jorge [16].

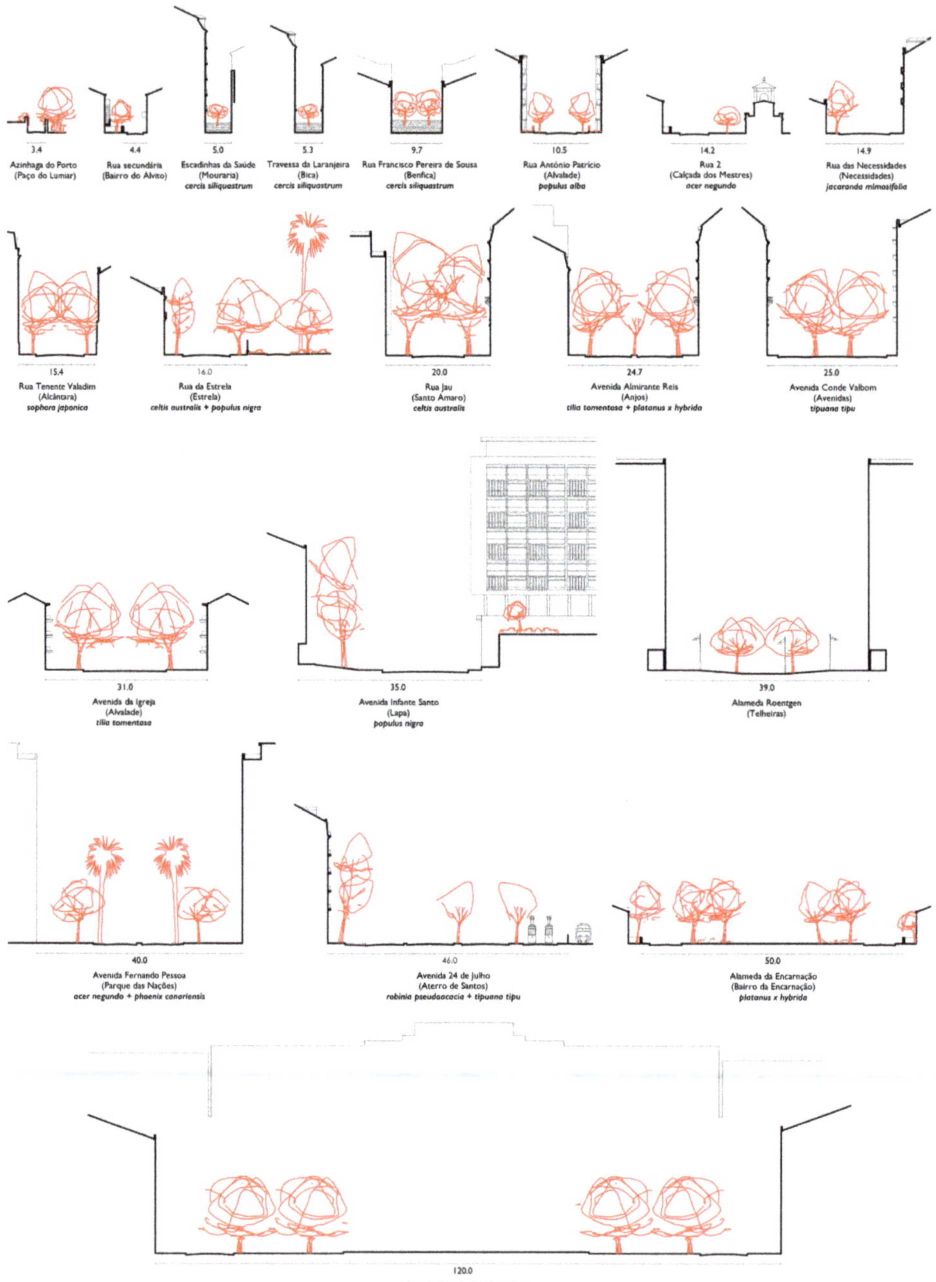



**Figure 3.** The diversity of formal relations between trees and street spaces in Lisbon.

In Lisbon, the presence of trees is significant and varied but was largely confined to the private courtyards within the urban blocks until the nineteenth century. At this time, the presence of arboreal elements in the public structure was meticulous, punctuating spaces of a more singular nature, as can still be observed in Alfama.

It was not until the mid-nineteenth century that trees began to play a truly essential role in the qualification and beautification of the city, especially in exceptional public spaces such as belvederes, squares or gardens, but also in streets of larger sections. This idea was supported by hygienist principles that favoured the healthiness of urban space, envisioned as broad, airy and with the presence of vegetation. This tendency quickly became a rule for both new and pre-existing spaces, in a logic that also became a hierarchy of the urban structure of the city [13].

The present form of the city clearly expresses this evolutionary process that we will here aim to synthesize by focusing on the analysis of different areas of Lisbon. The three chosen areas to be explored—Alfama, Avenidas and Alvalade—are sufficiently distinct in their form and context, allowing a comparative interpretation of the differences between their main morphological elements. Because of their historical and/or landscape context, each homogeneous area now presents a distinct physical structure with diverse urban functions and human activities. Furthermore, it is possible to recognize how different vegetation is composed and how the presence of vegetation influences the acknowledgement of the particular urban morphology of each area and of the city as a whole.

### 3.1. Alfama—the Tree as a Point

The urban fabric of Alfama corresponds to what remains of the old medieval nucleus of the city of Lisbon. This nucleus, also formed by the Baixa and Mouraria neighbourhoods, was until the 1755 earthquake circumscribed to the area within the Fernandina Wall. Alfama thus presents itself as a singular segment of this urban reality, representative of a way of producing the city that results from a gradual and organic process of urban fabric formation and transformation. In other words, it is the physical inheritance of the tension between different individual actions, desires and needs over time.

Developed in a clear adaptation to the topography, the urban fabric results in a composition where private and public spaces are easily recognizable and well defined. This means that urban blocks are defined by the urban layout and, at the same time, they define the boundary and alignment of streets, being densely occupied by housing buildings and occasionally containing gardens and open spaces inside.

The type of urban organization is the result of a way of developing the city that starts from the design of the street, where the urban block is the remaining element in the composition. The urban layout of Alfama is composed of a set of formal and toponymical diverse spaces: streets, alleys, stairways and arches. The structure that integrates them (Figure 4, above) develops mainly from two design logics, one that presupposes a desire to connect the river to the highest point of the slope in the most efficient way, and another, which perceives the singular elements, spaces and buildings, as fabric generators influencing the definition of their form and hierarchy.

Existing planted vegetation is scarce and encompasses mostly trees, such as *Phoenix canariensis, Tipuana tipo* or *Celtis australis*, taking form in sporadic positionings and mainly with a single tree element (Figure 5). Bearing in mind the topography of the place and the stature and volume that these trees may reach, they often constitute landmarks, contributing to orientation and location through their singularity. That is namely the case of the *Phoenix canariensis* that used to be at Largo de São Miguel (Figure 4, below), which died from a palm tree scarab (*Rhynchophorus ferrugineus*) but has been recently replaced with a younger specimen. This sporadic presence of isolated trees in the Alfama structure is also associated with secondary spaces such as bystreets, stairs or even in alleys.

There is also a great density of greenery in pots of various configurations and vegetation of different species that residents place at their front doors or windows, stairs or courtyards, appropriating public space. In some situations, it is possible to find climbing or suspended vegetation that partially covers handrails, arcades or buildings.

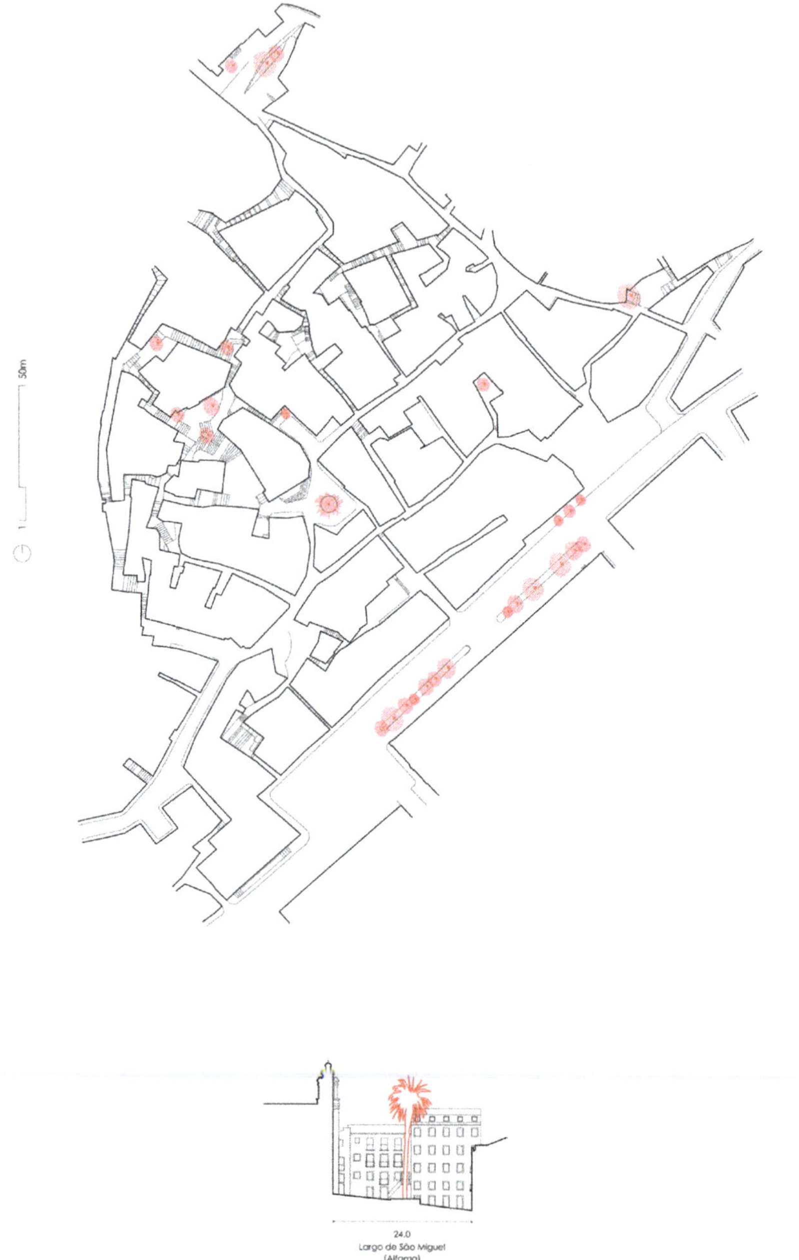


**Figure 4.** The sporadic and elemental incidences of the tree in the urban structure of Alfama (above). The case of Largo de São Miguel (below).

**Figure 5.** The diversity of singular trees in Alfama. Photographic source: Sérgio Proença (above) and Formaurbis LAB (below).

## 3.2. Avenidas—the Trees as a Line

The Avenidas is one of the richest urban and architectural environments of Lisbon, not so much by the representativeness of its buildings, but mainly due to an eclectic assortment of types and styles formed since the late nineteenth century by the Avenidas plan. Designed by Ressano Garcia, the plan led the development of the city to the north, using an elementary urban system composed of an orthogonal design matrix similar to what happened in other European cities of the nineteenth century. It clearly appears in response to the emergence of a new bourgeois class claiming a habitable, spacious and airy city with green spaces.

The plan is based on a juxtaposed structure of three different urban layouts organized around three central axis sequentially articulated along the valleys by squares. In addition, the proposal structure includes the integration of a set of pre-existing elements: some rural plots, buildings and, above all, a primary structure of roads connecting the city with the small rural nucleus beyond Lisbon [17].

The plan composition encompassed two opposite approaches when considering, on the one hand, the detail given to the design of public space and, on the other hand, significant uncertainty about private space by the absence of architectural regulations for buildings. These guiding principles, or the lack of them, have remained until the present day, forming an urban unit where public space remains practically untouched and the built fabric lives in constant transformation. In a built fabric formed mainly by housing typologies of different configurations, a new trend appeared from the 1970s onward to integrate services and commercial buildings into the main axes of the Avenidas. This process reinforces a tendency to concentrate human activities in the squares that articulate them.

In the genesis of the plan for Avenidas, Ressano Garcia and António Maria Avellar had the intention of planting different tree species in each avenue—*Platanus x hybrida* for Avenida da República, *Jacaranda mimosifolia* for Avenida 5 de Outubro, *Celtis australis* for Avenida Duque d'Ávila, *Tipuana tipu* for Avenida Conde Valbom, *Populus tremula* for Avenida Fontes Pereira de Melo, amongst others (Figure 6, right; Figure 7). This characteristic, that is still evident today, gives a singular character to each street, "the trees created the initial identity of the street" [13].

Plantations in this urban structure are mostly linear, emphasizing the rectilinear tracings of the different avenues (Figure 6, left). These linear structures, formed by the orderly distribution of trees along an axis, constitute alignments that develop parallel to the building that forms the street. Alignments allow the delimitation, separation and framing of elements and spaces that form the street cross-section, depending on their position. Usually, the alignments we find in the city of Lisbon are formed by two parallel lines of trees that are positioned on the sides of the street, separating the road from the pedestrian. This principle is mainly verified in the streets created between the late nineteenth and early twentieth centuries, yet we often find variations of streets with central alignments.

The case of Avenidas is a good example where the presence of these two types of alignments—lateral and central—is recurrent and balanced, such as the Alexandre Herculano and Filipe Folque streets (Figure 8). These two examples also allow the demonstration of different ways of forming these alignments, evidencing the existence of single lines and lines that can be multiplied within the same street space (Figure 9), usually composed of trees of the same species. Exceptionally, these alignments derive from tree line aggregations of different species, such as the case of Avenida da Liberdade (Figure 8), highlighting the structuring role of this street space in the context in which it is inserted. This example highlights the importance that alignments can have in the definition of a hierarchical urban structure, such as the case of Avenidas.

**Figure 6.** The regularity of tree alignments in the urban structure of Avenidas (left). The different tree species on each avenue (right).

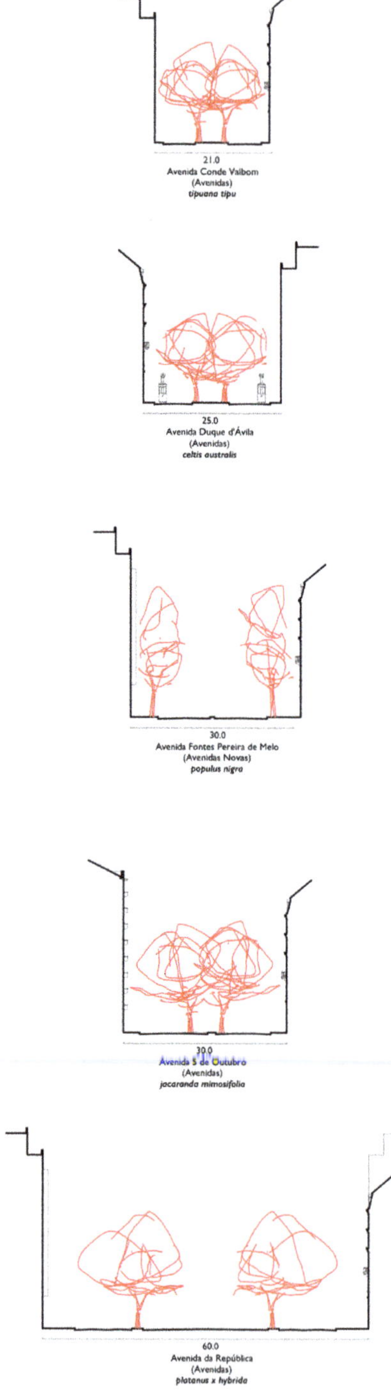



**Figure 7.** The diversity of formal relations between different tree species and the street space in Avenidas.

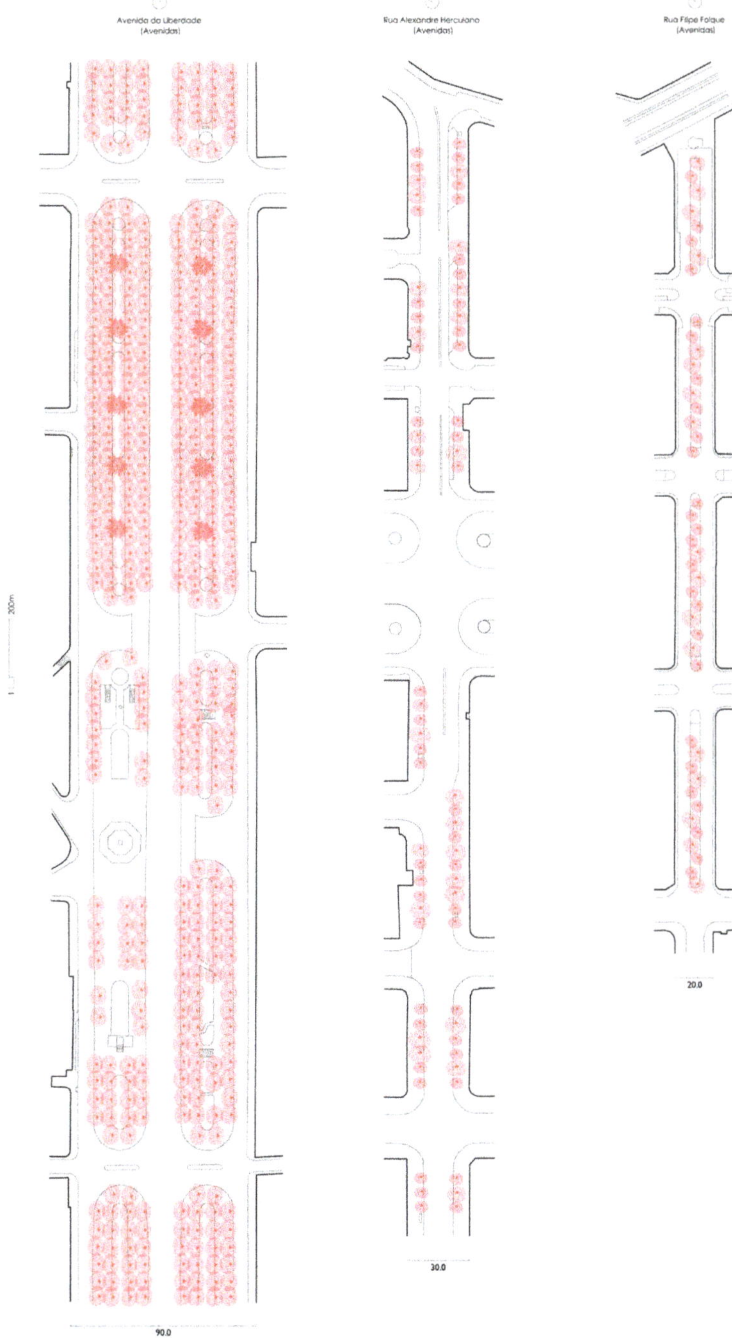



**Figure 8.** Three types of tree alignments in Avenidas (left to right): complex alignment—Avenida da Liberdade; simple and lateral alignment—Rua Alexandre Herculano; multiplied and central alignment—Rua Filipe Folque.

**Figure 9.** Two types of tree alignment in Avenidas: multiplied alignment—Avenida da Liberdade (above); simple alignment—Avenida Duque de Loulé (below). Photographic source: Formaurbis LAB.

### 3.3. Alvalade—the Trees as a Line vs. the Trees as a Volume

At a time when the housing issue was sensitive, given the social changes caused by World War II that changed the urban reality of the city, the Alvalade neighbourhood, designed by Faria da Costa in 1944, proved to be experimental and innovative in the way it was conceived and then executed. The urban plan was promoted by the municipality, integrated in one of the biggest transformation areas north of Lisbon already planned in the Master Plan of Étienne de Groer (1938–1948).

The implementation of the urbanization plan for the area south of Avenida Alferes Malheiro was managed and partially assured by the State, thus defining specific areas for private construction. This situation favoured not only the introduction of the neighbourhood unity principle, but also the coexistence of different typologies of social housing with significant variations in building and public space design, as the current urban fabric still reproduces [18]. Therefore, the design of the Alvalade neighbourhood results from an articulated and rich set of ideas of urban fabric production where the urban layout, plot structure and building are an integral part of the design. If we consider the principles governing the execution plan, we can easily observe that they remain in the current structure of the neighbourhood: the nature, form and hierarchy of the urban layout; the eight housing cells and the road organization of each one of them; the exceptional elements that serve as articulation and focus of the urban layout; central public spaces in each housing cell, associated with school equipment and green spaces; the preponderance and diversity of the housing typology; and the coexistence of different urban models and experiences [18] (Figure 8).

The Alvalade neighbourhood is today a qualified space and a reference of diverse urban environments in the context of the city of Lisbon. We can, therefore, find spaces of great intensity of flows and activities, located along the main axes, as well as spaces whose vocation is almost exclusive of access to the dwellings and, therefore, of smaller flow.

In accordance with the ideals planned for this area, vegetation is highly present and takes form in various ways (Figures 10 and 11). Indeed, several effects are combined through tree and shrub alignments as well as occasional volumes of dense irregular vegetation to provide not only a distinctive place with its own identity but also a place of proximity.

**Figure 10.** The complex incidence of trees in the urban structure of Alvalade.

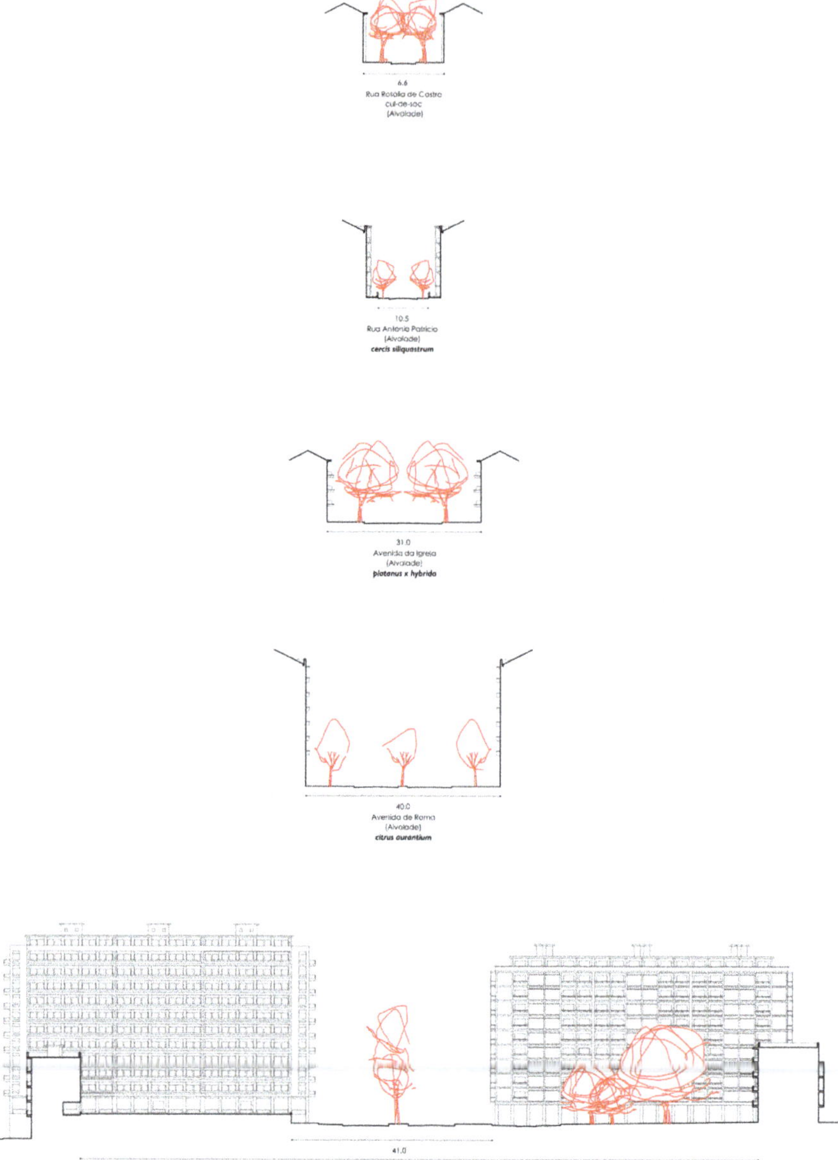

**Figure 11.** The diversity of formal relations between the tree and the street space in Alvalade.

One may note tree alignments, generally arranged for staging, emphasizing and framing a monument. That is namely the case of the alignments in Avenida da Igreja, developed in two parallel lines of trees (Figure 12, left). This alignment is reinforced by the presence of four *Cupressus semprevirens* in front of the church of São João de Brito. In other situations, namely in the "cul-de-sac" areas structured in "U's", vegetation is prominent (Figure 12, right). In these areas, tree alignments are combined with flower beds with all kinds of vegetation from shrubs to herbaceous and climbing plants. In this former situation, vegetation sets the boundaries of space, consolidating an urban identity through its textures. This example emphasizes that tree alignments are not always associated with streets of higher hierarchy. Indeed, in the case of Alvalade, tree alignments are identified in both structuring streets and housing impasses.

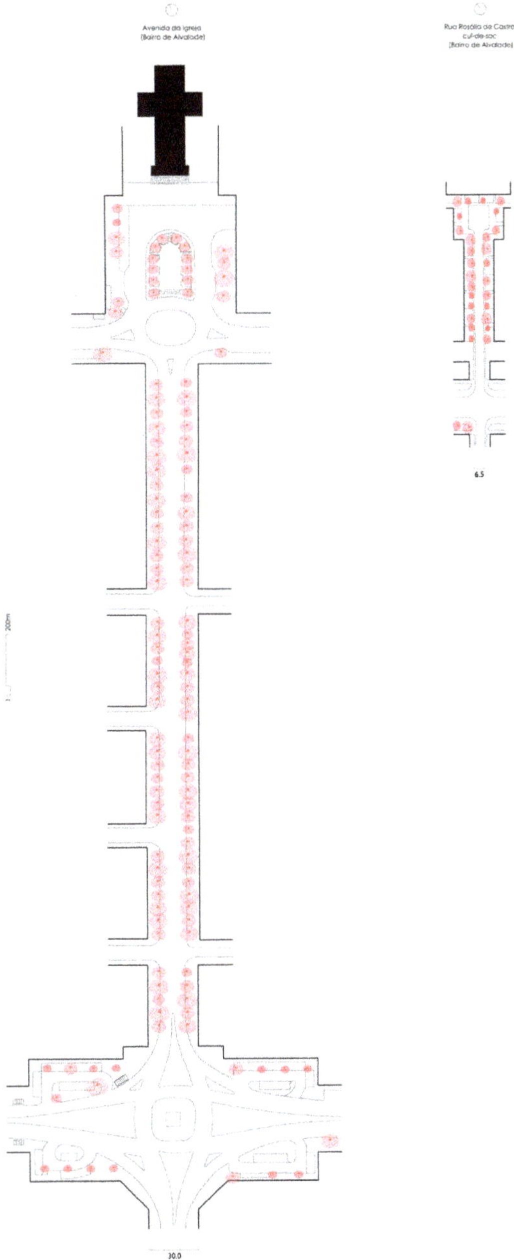


**Figure 12.** The tree alignments in Alvalade.

On the other hand, in the same neighbourhood, it is possible to identify aggregations of trees that adopt different languages, due to their more informal emplacements. In these cases, the trees are combined and arranged more organically and are often of distinct species, forming a dense green volume. In Alvalade, Bairro das Estacas and Avenida EUA (Figure 13), they assume a pioneering role in the city of Lisbon, as, amongst others, they are the first spaces that encompassed plantations of these characteristics, designed during the 1950s [19]. Today, it is possible to find other spaces in Alvalade of a more exceptional nature that adopt this system of trees aggregation in relatively dense volumes.

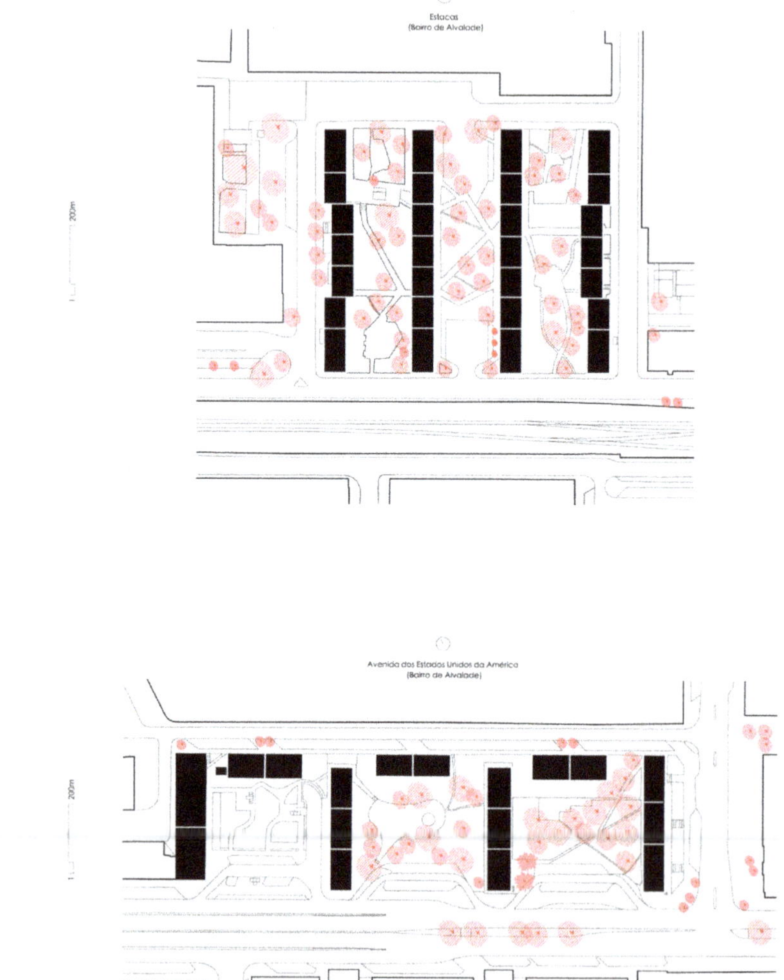


**Figure 13.** The trees as a volume in Alvalade.

Diversity may be the best word to describe Alvalade, not only in the multiple experiences of aggregation of built typologies but also of tree aggregation, of which the different tree alignments and the large green volumes of trees that fill the space between the buildings stand out unequivocally (Figure 14).

**Figure 14.** The trees as a line in Avenida da Igreja (above) vs the trees as a volume in Avenida EUA (below).

## 4. Conclusions: Vegetation as an Element of Urban Morphology

In the medieval morphology of Alfama, trees are scarce and appear in sporadic positionings. Most of these trees, individually, constitute landmarks. In some situations, groups of trees form protecting domes constituting semi-enclosed spaces and generally, throughout the homogeneous area, gatherings of vegetation in different types of pots adjoining the public space form part of the collective landscape. In the post-industrial Avenidas designed by Ressano Garcia, tree alignments are the dominant vegetation form, emphasizing the proposed rectilinear tracings. In some situations, the volume resulting from these alignments competes with the height of adjacent buildings. Finally, in the modern Alvalade neighbourhood, vegetation is highly present. It encompasses the combination of vegetation alignments with occasional volumes of dense and irregular vegetation. In some situations, regular tree and shrub alignments serve staging purposes. In other situations, trees and shrubs rather serve as borders that confine and consolidate privileged spaces.

In this study, the main characteristics of the urban structure are revealed, together with the importance of vegetation as an element of urban morphology (Figure 15), which significantly contributes to the understanding and interpretation of the city. If other homogeneous areas were analysed, new urban morphological components would very likely be identified.

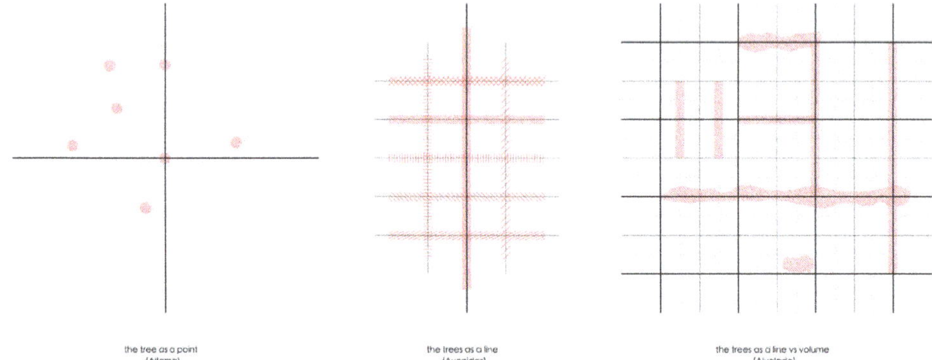



**Figure 15.** Systematization of the tree aggregation forms in accordance with the analyzed urban structures.

Most importantly, this initial systematization process serves to improve knowledge on the reading of urban greenery, also offering another set of values for contemporary design processes. Indeed, the use of morphological analysis as an analytical methodology is both an exploratory tool to understand existing urban, architectural and vegetation forms, and a tool for the creation of new forms. In this sense, the urban morphology exercise gains relevance in architectural and landscape architectural education, providing tools to improve the process of creative research and design. Overall, these outputs, methodology and theoretical framework aim to contribute support to a design process that can respond to the challenges of the contemporary city and improve human well-being through the acknowledgement of vegetation as a contributor to the creation of specific, unique and unrepeatable spaces within urban landscapes.

**Author Contributions:** R.J., M.M.S. declare that each have contributed equally to the production of this document. All authors have read and agreed to the published version of the manuscript.

**Funding:** This research was supported by Fundação para a Ciência e a Tecnologia through a PhD fellowship in Urbanism (SFRH/BD/116371/2016).

**Acknowledgments:** The authors would like to thank Formaurbis Lab, a research group on urban morphology from the Lisbon School of Architecture—Universidade de Lisbon and especially Sérgio Proença, a team member, for allowing full access to the data of Lisbon: a set of elements used as a basis for the production of the manuscript drawings and some photographs.

**Conflicts of Interest:** The authors declare no conflict of interest.

## References

1. Sitte, C. *L'art de batir les villes, L'urbanisme Selon ses Fondements Artitiques*; Editions du Seuil: Paris, France, 1996.
2. Stübben, J. La Construcion des Villes, Régles Pratiques et Ethétiques a Suivre pour l'Élaboration de Plans de Villes. In *Rapport présenté au Congrès International des Ingènieurs de Chicago*; E. Lyon-Claesen: Bruxelas, Belgium, 1985.
3. Coelho, C.; Costa, P.; Leite, J.; Silva, J.; Trindade, L.; Pereira, P.; Proença, S.; Fernandes, S.; Monteys, X. *Cadernos de Morfologia Urbana. Estudos da cidade portuguesa*; Argumentum: Lisboa, Portugal, 2015.
4. Moudon, A. The changing morphology of suburban neighborhoods. In *Typological Process and Design Theory*; Petruccioli, A., Ed.; Aga khan Program for Islamic Architecture: Cambridge, MA, USA, 1998; pp. 141–157.
5. Leite, J.; Justo, R. Typo-morphology: From research to architectural education. In *Architectural Research Addressing Societal Challenges, Proceedings of the EAAE ARCC 10th International Conference, Lisbon, Portugal, 15–18 June 2016*; da Costa, M., Roseta, F., Lages, J., da Costa, S., Eds.; Taylor & Francis Group: London, UK, 2017; pp. 1175–1190.
6. Aymonimo, C. El estudio de los fenómenos urbanos. In *Análisis Urbano*; Barajas, P.Y., Ed.; Instituto Universitario de Ciencias de la Construcción: Sevilha, Espanha, 1966.

7.   Lamas, J. *Morfologia Urbana e desenha da cidade*; Fundação Calouste Gulbenkian: Lisboa, Portugal, 2004.
8.   Lynch, K. *A imagem da cidade*; Edições 70: Lisboa, Portugal, 1996.
9.   Cullen, G. *Paisagem Urbana*; Edições 70: Lisboa, Portugal, 2004.
10.  Kostof, S. *The City Shaped: Urban Patterns and Meanings Through History*; Thames & Hudson: London, UK, 1991.
11.  Mailliet, L.; Bourgery, C. *L'arboriculture Urbaine*; IDF: Paris, France, 1993.
12.  Stefulesco, C. *L'urbanisme Vegetal*; Institut pour le Development Forestier: Paris, France, 1993.
13.  Proença, S. A Diversidade da rua na Cidade de Lisboa. Morfologia e Morfogénese. Ph.D. Thesis, Lisbon School of Architecture, Universidade de Lisboa, Lisboa, Portugal, 2014.
14.  Cabral, F.; Telles, G. *A Árvore em Portugal*; Assírio & Alvim: Lisboa, Portugal, 1999.
15.  Saraiva, G.; Almeida, A. *Árvores na Cidade. Roteiro das Árvores classificadas de Lisboa*; By the Book: Lisboa, Portugal, 2016.
16.  Jorge, F. *Lisboa vista do céu*; Argumentum: Lisboa, Portugal, 2013.
17.  Justo, R. O Diacronismo do Tecido. In *O Tempo e a Forma*; Coelho, C., Ed.; Argumentum: Lisboa, Portugal, 2018; Volume 2, pp. 50–69.
18.  Costa, P. *O Bairro de Alvalade. Um paradigma no Urbanismo Português*; Livros Horizonte: Lisboa, Portugal, 2010.
19.  Tostões, A. *Os verdes anos na arquitectura portuguesa dos anos 50*; Faculdade de Arquitectura da Universidade do Porto: Porto, Portugal, 1997.

*Article*

# Mapping and Analyzing the Park Cooling Effect on Urban Heat Island in an Expanding City: A Case Study in Zhengzhou City, China

**Huawei Li [1,2,*], Guifang Wang [1], Guohang Tian [2] and Sándor Jombach [1]**

[1]  Department of Landscape Planning and Regional Development, Faculty of Landscape Architecture and Urbanism, Szent István University, 1108 Budapest, Hungary; guifang.wang@phd.uni-szie.hu (G.W.); jombach.sandor@tajk.szie.hu (S.J.)

[2]  Department of Landscape Architecture, College of Forestry, Henan Agricultural University, Zhengzhou 450002, China; tgh@henau.edu.cn

*  Correspondence: li.huawei@phd.uni-szie.hu

Received: 29 December 2019; Accepted: 11 February 2020; Published: 14 February 2020

**Abstract:** The Urban Heat Island (UHI) effect has been extensively studied as a global issue. The urbanization process has been proved to be the main reason for this phenomenon. Over the past 20 years, the built-up area of Zhengzhou city has grown five times larger, and the UHI effect has become increasingly pressing for the city's inhabitants. Therefore, mitigating the UHI effect is an important research focus of the expanding capital city of the Henan province. In this study, the Landsat 8 image of July 2019 was selected from Landsat collection to obtain Land Surface Temperature (LST) by using Radiative Transfer Equation (RTE) method, and present land cover information by using spectral indices. Additionally, high-resolution Google Earth images were used to select 123 parks, grouped in five categories, to explore the impact factors on park cooling effect. Park Cooling Intensity (PCI) has been chosen as an indicator of the park cooling effect which will quantify its relation to park patch metrics. The results show that: (1) Among the five studied park types, the theme park category has the largest cooling effect while the linear park category has the lowest cooling effect; (2) The mean park LST and PCI of the samples are positively correlated with the Fractional Vegetation Cover (FVC) and with Normalized Difference Water Index (NDWI), but these are negatively correlated with the Normalized Difference Impervious Surface Index (NDISI). We can suppose that the increase of vegetation cover rate within water areas as well as the decrease of impervious surface in landscape planning and design will make future parks colder. (3) There is a correlation between the PCI and the park characteristics. The UHI effect could be mitigated by increasing of park size and reducing park fractal dimension (Frac_Dim) and perimeter-area ratio (Patario). (4) The PCI is influenced by the park itself and its surrounding area. These results will provide an important reference for future urban planning and urban park design to mitigate the urban heat island effect.

**Keywords:** park cooling effect; park characteristic; urban heat island; land surface temperature; Zhengzhou; expanding city

---

## 1. Introduction

The Urban Heat Island (UHI) problem has been studied for more than 200 years since it was discovered in 1818 [1]. It refers to the phenomenon that the temperature of urban areas is higher than the temperature of its surrounding area [2–4]. The main reason for this phenomenon is the process of urbanization, as the vegetation is replaced with the built-up area during urban development [5]. This process leads to changes in the physical properties of the surface structure which modifies the thermal environment of urban areas. At present, urbanization is a major driving force within developing countries and their rapid urban growth towards becoming a developed nation. Taking China as an example, according to the National Bureau of Statistics of China, the urbanization rate in China was 59.58% in 2018. This number is 12.58% higher than in 2008. This high-speed urbanization had led to widespread UHI problems in China [6]. With this rapid increase in urbanization has come an increase in energy consumption, with individuals attempting to reduce these adverse temperature effects (e.g., air conditioning, private vehicle usage). As a consequence, higher energy consumption based on coal power plants may result in air pollution, water pollution, and additional climate changes. More specifically, UHI compromised the health and life quality of citizens [7].

Generally speaking, UHI can be measured by two methods: one of them is Surface UHI (SUHI); the other one is atmospheric UHI. Atmospheric UHI is defined into two different types: canopy layer UHI and boundary layer UHI [8,9]. From studies on the UHI effect, SUHI is widely used to characterize the UHI effect in the case of regional or urban studies. Due to the development of remote sensing technology, a high number of studies used satellite imagery to derive land surface temperature (LST) [6,10–12]. Concerning the investigation of UHI characteristics and changes in large-scale, these studies mainly focus on: (1) The spatial distribution of UHI; (2) the methods of satellite image inversion; (3) the relationship between land use land cover (LULC) and LST [13]. Furthermore, landscape pattern analysis in a regional scale was also proved to be a proper method in UHI research [14].

However, in small-scale UHI studies, the UHI is mainly characterized by the actual measured air temperature [15–17]. These kinds of studies mainly focused on the temperature difference between the green space and other land types, and the method of characterizing UHI intensity. In addition, some studies add microclimate factors and use the Local Climate Zone (LCZ) factors [3,18] to investigate UHI. These microclimate conditions such as wind speed, wind direction, humidity, solar light intensity, surface reflectance and other localized effects on temperature [19–21]. Those studies showed that the green space cools the air due to the transpiration of the plants, which contributes to low UHI. In addition, local wind speed and wind direction also modify air temperature. The higher the surface albedo, the lower the temperature is found to be [22]. Besides the complexity of local climate and related environmental conditions the measurement of air temperature is limited by the monitoring system, including pieces of equipment, experts, method and such factors. The accuracy of data collection is always a key factor and it is almost impossible to conduct ideal UHI research on a wide range of space-time scales. These data sources are however very useful in understanding the generalized sources of data studied.

In the studies of Urban Cold Island (UCI) effects and UHI mitigation [23,24], two methods were used to quantify the cooling effect of green space and park areas. These are called Green space Cooling Intensity (GCI) and Park Cooling Intensity (PCI) [25,26]. GCI is defined as the temperature difference between green space and the average temperature of the whole study area. While the PCI usually determined as the temperature difference between the inside park area and its outside within a 500 m buffer area [27,28]. These two methods are used to describe the cooling effect of green spaces and parks. Studies also show that vegetation coverage has a significant effect on the reduction of UHI [10,29,30]. On top of this, the impervious surface area is positively correlated with LST and contributes the most to UHI [10,31,32]. Some studies use landscape factors to analyze the UCI and GCI on mitigating the UHI effect [33–35]. The parameters include shape index (Shape_Idx), fractal dimension (Frac_Dim) and landscape connectivity. One research study focused on the role of green space in reducing the UHI effect. This study looked at the distance changes of the green space cooling effect in relation to the characteristics of green space bodies with regards to size, perimeter, shape index, fractal demission, and UHI. The results of these studies showed that the cooling effect of green space is complex [25–27]. PCI related studies mostly used remote sensing imagery as base data. According to a study based partly on surrounding vegetation, water body and impervious surface [26], the cooling effect of parks is also depending on types of outside the park spaces. The air temperature is also employed to examine the cooling effect of parks on UHI [36,37]. It is also noted that the patch and pattern of a park has a relationship to the cooling effects that it has on the UHI it forms part of. Some researches employ the thermal conduct theory, a physical science methodology, to investigate the heat balance (between park and its surrounding area) when studying the urban thermal environment [38–40], this method of study can assist in providing a methodical approach to understanding the cooling effect of park.

At present, cities have grown exponentially. The actual scale and speed of China's overall urbanization process has never seen before in modern urban development. Super-large cities and megacities have become commonplace in China's inland areas. These big cities result in urban environment problems like UHI, For example, In Zhengzhou city, Studies showed that from 1996 to 2014, the average LST increased by 2.94 °C in Zhengzhou city [41], and the UHI change was positively correlated with land cover changes over this period. It has also been proven that the increase of the built-up urban areas [42] showed a negative correlation with the vegetation cover rate in Zhengzhou. UHI studies in the city of Zhengzhou analyzed many factors, but these neglected to focus on green space categories types while also not using large enough sample sizes.

In this paper, we selected the latest cloud-free satellite image acquired on 07 July 2019 as base data to focus on the UHI effect characteristics in the megacity of Zhengzhou. Choosing 123 parks as samples identified though high-resolution Google Earth images, we investigated the cooling effect characteristics of the chosen parks. The cooling intensity and park buffer sizes were studied, and the correlation between the park patch metrics and the cooling intensity was explored. We aimed to: (1) Analyze the PCI differences among five park types; (2) analyze the park LST and its relation to vegetation, water surface area and impervious surface factors; (3) analyze park LST and its relation to park patch indices; (4) analyze PCI and its relation to park patch indices and impact factors of park surrounding areas. As a whole, this research was conducted to analyze the relationship between park cooling effect and its related impact factors, to understand UHI characteristics in Zhengzhou. The intention of this analyses understanding is to give guidance for stakeholders, as well as to the developers of urban planning strategies to address UHI.

## 2. Study Area

Zhengzhou (34°16′–34°58′N, 112°42′–114°14′E) is the capital city of Henan Province in Central China. It is south of the North China Plain and the Yellow River (Figure 1). It is one of the largest transportation hubs in China. The population of the city was approximately 9.56 million, according to the 2017 census [43]. The population density is the second highest in China. Zhengzhou lies in the north warm-temperate zone, characterized by a warm climate, because of this it has four distinct seasons, characterized by a dry spring (March–May), and a hot and rainy summer (June–September).

Urbanization was taken place at a rapid rate in the past few decades in Zhengzhou city. This makes it an ideal sample city to understand the problems which are seen across urbanized cities across the world. Population growth and government level development policy is the main driver for the dramatic urban expansion. For example, the population of Zhengzhou increased from 4.2 Million in 1978 to 9.9 million in 2017, which means more than 100% growth in 40 years [44]. At the same time, Zhengzhou was designated as the core city of "Central Plain Economic Zone" and "Central Plains Urban Agglomeration." In 2016, Zhengzhou was officially named as the eighth "National Central City" in 2017 by the central government in China. This state-level policy provides a significant number of opportunities for the development of Zhengzhou.

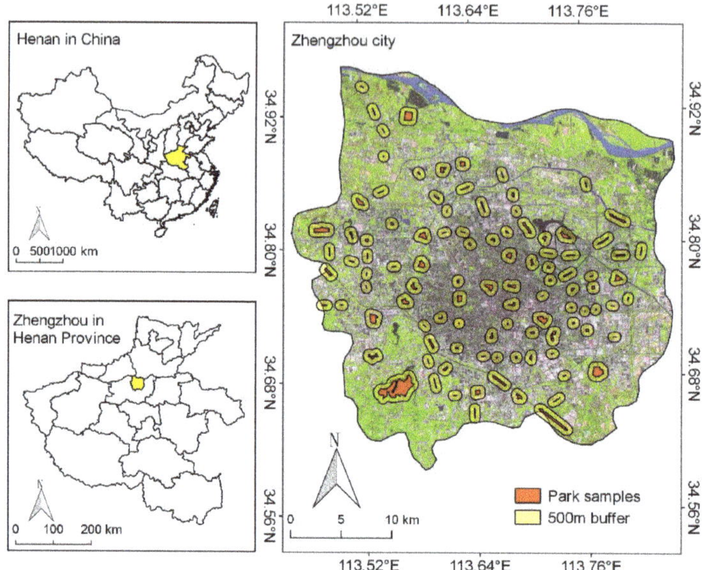


**Figure 1.** Location of the study area Zhengzhou, Henan province in China and 123 park samples in Zhengzhou.

## 3. Data Sources and Methods

### 3.1. Data Used

In this work, one satellite image from USGS (earthexplorer.usgs.gov; Table 1) was used to extract Land Surface Temperature (LST) in Zhengzhou city. The Landsat 8 image was selected from a collection of summer images obtained between May and September in an attempt best considering the LCZ of the study area. It is cloud-free and has high quality with a resolution up to 30 meters. In conjunction with the satellite image, we also employed additional base maps obtained from high-resolution Google Earth images, low altitude UAV images, and the official urban land use map of Zhengzhou city.

**Table 1.** Characteristics of the Landsat-8 OLI image used in this study.

| Date | Path/Row | Band | Wavelength (μm) | Resolution (m) |
|---|---|---|---|---|
| 07 July 2019 | 124/36 | Band 1—Ultra Blue | 0.435–0.451 | 30 |
| | | Band 2—Blue (B) | 0.452–0.512 | 30 |
| | | Band 3—Green (G) | 0.533–0.590 | 30 |
| | | Band 4—Red (R) | 0.636–0.673 | 30 |
| | | Band 5—Near Infrared (NIR) | 0.851–0.879 | 30 |
| | | Band 6—Shortwave Infrared (SWIR) 1 | 1.566–1.651 | 30 |
| | | Band 7—Shortwave Infrared (SWIR) 2 | 2.107–2.294 | 30 |
| | | Band 10 *—Thermal Infrared (TIR) | 10.60–11.19 | 100 * (30) |

* Native resolution was 100 m, TIRS thermal constant of band 10: $K_1 = 774.89$; $K_2 = 1321.08$.

### 3.2. Retrieval of LST and the Average LST Calculation

The radiative transfer equation (RTE) method of land surface temperature (LST) has been widely recognized and generally divided into five steps [45–48]:

(1)  Conversion to Spectral Radiance [49];

(2)  Conversion to top of Atmosphere Radiance [49];

(3)  Conversion to Top of Atmosphere Brightness Temperature [49];

(4)  Calculation of Proportion of Vegetation [50];
(5)  Estimation of estimate land surface emissivity (LSE) [48];
(6)  Retrieval of land surface temperature (LST).

### 3.3. Sample Selection

Based on the classification applied in Chinese urban planning regulations and the distribution of parks in Zhengzhou city, this paper selected five functional types of parks, 123 (one hundred and twenty three) parks in total as study sites (Table 2). The parks boundaries were determined based on high-resolution Google Earth images and low-altitude UAV (drone) images. As the original spatial resolution of the Landsat thermal infrared band is 100 m, we selected sample parks larger than 2 hectares.

**Table 2.** Statistics and details of 123 sample parks by types.

| Types | Number | Percentage | Maximal Area (ha) | Minimal Area (ha) | Main Example |
|---|---|---|---|---|---|
| Urban Park | 59 | 48% | 87.16 | 3.12 | City park, District park |
| Theme Park | 10 | 8% | 108.53 | 10.61 | Botanical garden, Zoo |
| Street Park | 28 | 23% | 25.26 | 1.23 | Pocket park, community park. |
| Linear Park | 19 | 15% | 62.02 | 2.13 | Riverside park, roadside park |
| Urban Square | 7 | 6% | 6.64 | 2.15 | Square |

### 3.4. Determination of the Park Cooling Intensity (PCI)

Park Cooling Intensity (PCI) usually calculates the temperature difference between the inside and outside of the park [28,51,52]. It can be air temperature or land surface temperature. In this study, the PCI (units in °C) was defined as the mean LST difference. Equation (1):

$$PCI = \Delta T = T_u - T_p \tag{1}$$

where $T_u$ is the mean LST of an urban area of the 500m buffer zone outside of the park, and $T_p$ is the mean LST inside the park. The buffer zone includes the area around the park, which contains different land cover types: Buildings, roads, impervious surfaces, trees, and green spaces.

### 3.5. Patch Descriptors of the Park

In this paper, several indicators were applied to characterize the impact factors on PCI (Table 3). By using the ArcGIS 10.2 tools, we calculated fractal dimension (Frac_Dim), perimeter-area ratio (Patario) and shape index (Shape_ldx) in patch level. From the previous studies, those three indicators were used as the main patch metrics, and had been widely employed to analyze landscape patterns, both in class level and patch level [33–35]. These initial base studies were successful in demonstrating the characteristics of landscape patterns both in regional and local scale [53]. Here we investigated the relation of these indicators to park cooling effect in sample areas of Zhengzhou city. Low fractal dimension is described as simple, non-waving, straight boundaries, high fractal dimension means waving edges of park and surroundings.

**Table 3.** Park metrics used in this study.

| Name | Equation | Description |
| --- | --- | --- |
| **Perimeter-Area Ratio *** | $\text{Paratio} = \frac{Pi}{Ai}$ | $P_i$ = perimeter (m) of patch *i*. $A_i$ = area (m$^2$) of patch *i*. **Paratio** equals the ratio of the patch perimeter (m) to area (m$^2$) [54]. |
| **Landscape Shape Index *** | $\text{Shape\_Idx} = \frac{0.25 Pi}{\sqrt{Ai}}$ | **Landscape shape index** provides a standardized measure of total edge or edge density that adjusts for the size of the landscape [54]. |
| **Fractal Dimension Index *** | $\text{Frac\_Dim} = \frac{2\ln(0.25 Pi)}{\ln Ai}$ | **Fractal Dimension Index** reflects the extent of shape complexity across a range of spatial scales [54]. |

Notes: * Source: FRAGSTATS: https://www.umass.edu/landeco/research/fragstats/documents/fragstats_documents.html.

In addition to the three indicators, we also used three indices to classify the satellite image of the study area (Table 4), which have been successfully proven by other researchers. The three indicators are Normalized Difference Water Index (NDWI) [55], Fractional Vegetation Cover (FVC) [56] and Normalized Difference Impervious Surface Index (NDISI) [57]. These indices can represent the surface coverage condition inside of the park.

**Table 4.** Spectral indexes of Landsat image used in this study.

| Name | Equation | Description |
| --- | --- | --- |
| **NDWI** | $\text{NDWI} = \frac{NIR-SWIR}{NIR+SWIR}$ | **Normalize Difference Water Index (NDWI)** is a remote sensing based indicator sensitive to the open water surface and water content of leaves [55]. |
| **NDVI** | $\text{NDVI} = \frac{NIR-R}{NIR+R}$ | **Normalize Difference Vegetation Index (NDVI)** is used o determine the density of green on a patch of land [10]. |
| **FVC *** | $\text{FVC} = \frac{NDVIi-NDVImin}{NDVImax-NDVImin}$ | **The Fractional Vegetation Cover (FVC)** is mainly depicts the vegetation abundance of ground surface [58]. |
| **MNDWI** | $\text{MNDWI} = \frac{G-SWIR}{G+SWIR}$ | **Modified Normalize Difference Water Index (MNDWI)** is an indicator used to determine the open water area [59]. |
| **NDISI** | $\text{NDISI} = \frac{TIR-[(MNDWI+NIR+SWIR)/3]}{TIR+[(MNDWI+NIR+SWIR)/3]}$ | **Normalize Difference Impervious Surface Index (NDISI)** indicator is used to estimate impervious surface [57]. |

Notes: * Where NDVIi is NDVI value of a pixel i; NDVImin is minimum value; NDVImax is maximum value. R, G, NIR, SWIR, TIR, (Table 1).

### 3.6. Analysis Methods

Statistical analysis was performed by SPSS 25.0 and Microsoft Excel. After retrieval of the LST, FVC, NDWI, NDISI values from the satellite image, QGIS was used to obtain summarized values of each sample area. Then SPSS was applied to conduct the linear regression analysis to quantify the relationship among LST, FVC, NDWI, NDISI, and PCI. For the park patch metrics calculation, we used the ArcGIS spatial analysis method to obtain the following parameters of each sample park: area, paratio, shape index and fractal dimension. The same linear regression analysis was made to PCI and LST. Additionally, the related coefficient was also utilized to detect and verify the result.

For the regression analyses, first, we use Pearson correlation analysis to obtain the main significant impact factors, and then analyze the regression relationship between the two factors in a targeted manner to find the optimal curve fitting model. The final presented fitting model (Figures 3–6) is the best explanation of the relationship between specific factors within the selected sample park.

## 4. Results

### *4.1. Relation between Park Types, LST and PCI*

Following the method in Section 3.2, the LST map based on the 07 July 2019 satellite image was derived (Figure 2a). Generally, the mean LST of the parks was lower than the mean LST of Zhengzhou city. We analyzed the PCI of all the selected samples by comparing the five park types (Table 2), and results show that the PCI of five park types were different (Table 5): The average temperature of the theme park category is 30.01 °C; noticeably it is 2.14 °C lower than the average temperature of Zhengzhou city. Its cooling effect is the strongest, where the average PCI reached 2.76 °C. The urban squares had the highest temperature, with an average LST of 32.13 °C, which is still 0.02 °C lower than the average LST in Zhengzhou.

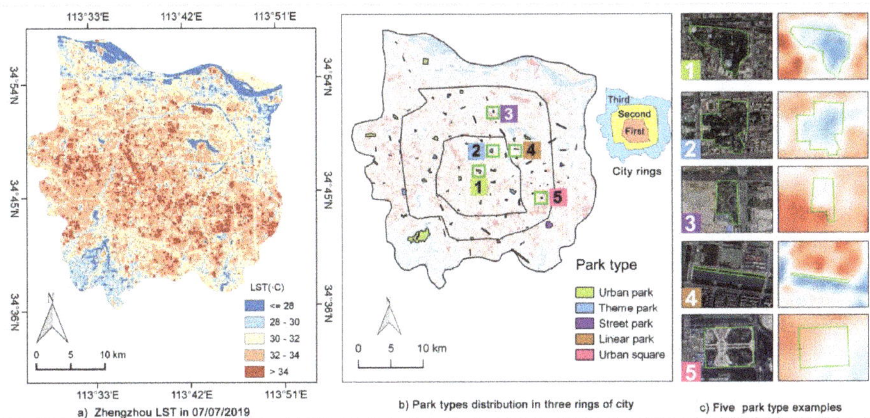


**Figure 2.** (**a**) LST map of Zhengzhou city in 07 July 2019; (**b**) Distribution of 123 parks samples; (**c**) Five park type examples: 1-urban park; 2-theme park; 3-street park; 4-linear park; 5-urban square.

**Table 5.** Statistics of LST and average PCI in different park types.

| Code | Type | Number | Mean LST (°C) | Max (°C) | Min (°C) | Average PCI (°C) |
|---|---|---|---|---|---|---|
| 1 | Urban park | 59 | 30.43 | 33.63 | 27.63 | 1.71 |
| 2 | Theme park | 10 | 30.01 | 34.62 | 25.97 | 2.76 |
| 3 | Street park | 28 | 31.32 | 37.80 | 25.34 | 0.8 |
| 4 | Linear park | 19 | 31.47 | 35.42 | 28.56 | 0.64 |
| 5 | Urban square | 7 | 32.13 | 33.90 | 31.10 | 1.44 |
| 6 | Zhengzhou city | - | 32.15 | 46.09 | 20.11 | - |

The result (Table 5) showed that PCI of the theme park category was the largest, this is related to the content of the theme park. Linear parks had the lowest cooling effect; the PCI value was only 0.64 °C. Meanwhile, the cooling effect of the street park category is also at a comparatively low level, with its PCI only 0.16 °C higher than the PCI of the street park group. The PCI of the urban square category was in the middle, it reached 1.44 °C, but its average LST was the hottest among the five types. So based on the results, we can conclude that theme parks have the most substantial cooling effect in Zhengzhou city, while the linear park category contributes with a less cooling effect.

### *4.2. Relation between Park LST and Its Impact Factor*

First, we analyzed the relation between park LST and the spectral indices inside the park. The results showed that the mean park LST was significantly related to the FVC, the NDISI, and the NDWI (Figure 3). The cooling effect of the park is directly proportional to the park's vegetation percentage ($R^2$ = 0.489), indicating that more vegetation cover makes parks cooler (Figure 3c). For example, Xiongerhe park's FVC has one of the highest values (0.408), while the mean temperature is the lowest (28.06 °C). Moreover, the average PCI of Xiongerhe shows it is much colder (2.18 °C) than its surrounding area. The results showed that the NDWI plays a negative role in park LST (Figure 3b), indicating that the NDWI value strengthens the cooling effect of the park. On the contrary, the NDISI has a relatively positive effect on park LST. From the regression model between LST and NDISI, the coefficient of determination ($R^2$) reached 0.926 (Figure 3c), revealing that the impervious surfaces have a significant impact on park temperature. The impervious surface is the main contributor to warm conditions of parks. We can conclude that water and vegetation have a positive impact on park cooling roles in Zhengzhou while the impervious surface increases the park warmth.

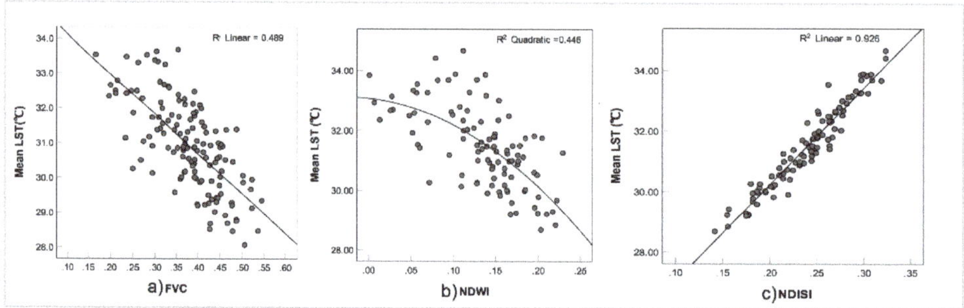

**Figure 3.** Regression analysis among mean park LST and, (**a**) FVC, (**b**) NDWI, (**c**) NDISI.

Secondly, we analyzed the relation between park LST and park characteristics (patch metrics). The result of the analysis shows that patch metrics have relations to park LST. From Figure 4, park size is negatively correlated with the mean park LST (Figure 4a, $R^2$ = 0.308), which means the park size is one of the main factors of LST. We can see from the Figure 4a, if the park size was larger than 40 ha, the average LST was below 31 °C, and the average LST appeared in a wide temperature range among the parks with size below 20 ha. Fractal dimension (Frac_Dim) and perimeter area ratio (Paratio) show a positive correlation with the park LST, and the coefficient of determination $R^2$ is 0.191, 0.280. This indicates that these two factors also have an impact on LST. The shape index has no significant correlation with park LST (Figure 4d). For example, the park with a maximum shape index (2.13) has the same LST (28.80 °C) as parks with the lowest shape index (1.21) (Figure 4e). From the results, we can conclude that the park size and perimeter-area ratio play a more critical role than other patch metrics in the sample parks of Zhengzhou city.

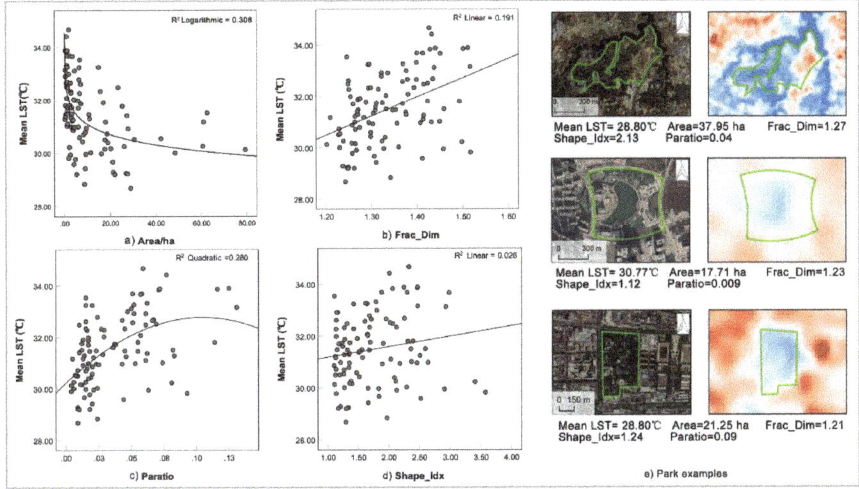

**Figure 4.** Regression analysis among mean park LST and park characteristics: (**a**) Size; (**b**) Frac_Dim; (**c**) Paratio; (**d**) Shape_Idx; (**e**) Three selected park examples.

### 4.3. Relation between PCI and Its Impact Factor

First, we analyzed the relation between PCI and the spectral indices inside the park. The results of correlation with PCI are shown in Figure 5. We found that FVC has a positive effect on PCI: the more FVC we have, the higher PCI appears. However, the coefficient of determination $R^2$ is only 0.237. This means that PCI only partly depends on vegetation cover. Figure 5b indicates that higher NDWI contributes to higher PCI, this quadratic regression analysis coefficient of determination ($R^2$) is 0.433. Among the three factors, the NDISI has the strongest relationship with PCI (Figure 5c), the coefficient ($R^2$) is 0.618, which means the impervious surface has a significant influence on PCI. So, from the park spectral indices results, we can recognize the park vegetation and water percentage play a decisive role in PCI, while the high impervious surface reduces the cooling effect of parks.

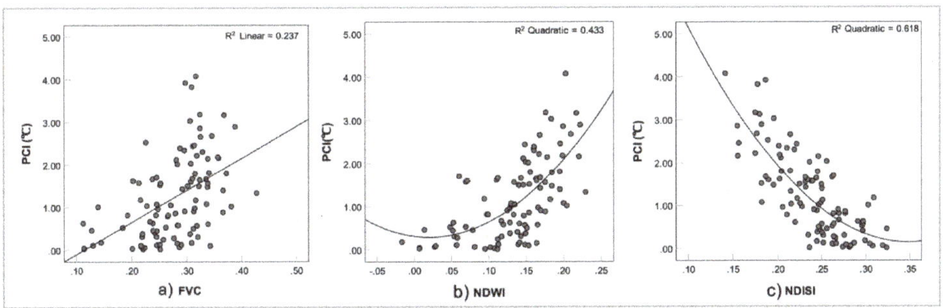

**Figure 5.** Regression analysis of park PCI and (**a**) mean FVC; (**b**) mean NDWI; (**c**) mean NDISI.

Secondly, we analyzed the relation between PCI and park characteristics (patch metrics). PCI has a complex correlation with park patch metrics (Figure 6). Among the four analysis results, the size, fractal dimension (Frac_Dim), and perimeter area ratio (Paratio) regression coefficient of determination $R^2$ is 0.321, 0.355, 0.439, respectively (Figure 6a–c) which means those three factors contribute to the PCI in general. While the shape index showed no significant correlation, as its linear model $R^2$ is 0.089 (Figure 6d). Park shape index does not contribute to PCI among the selected sample parks of Zhengzhou.

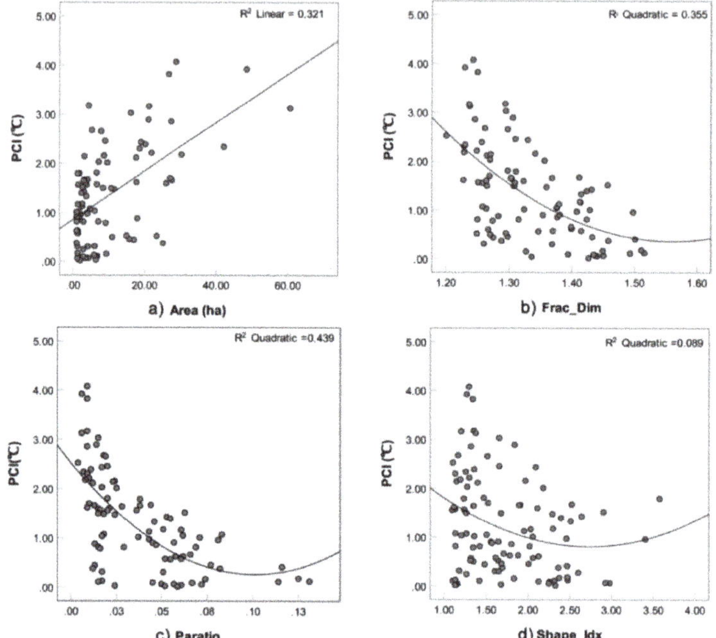

**Figure 6.** Regression analysis among PCI and mean park characteristics: (**a**) Size; (**b**) Frac_Dim; (**c**) Paratio; (**d**) Shape_Indix.

Thirdly, we investigated the relationship between PCI and the impact factor of the park surrounding area. In order to analyze the impact of PCI and the type of land cover around the parks, we selected 43 parks with similar mean LST (within the range of 29.0 °C–30.0 °C) from 123 samples (Figure 7). For external land cover types, we use spectral indicators: NDVI, NDWI, NDISI to measure vegetation, water coverage and impervious surfaces of the surroundings.

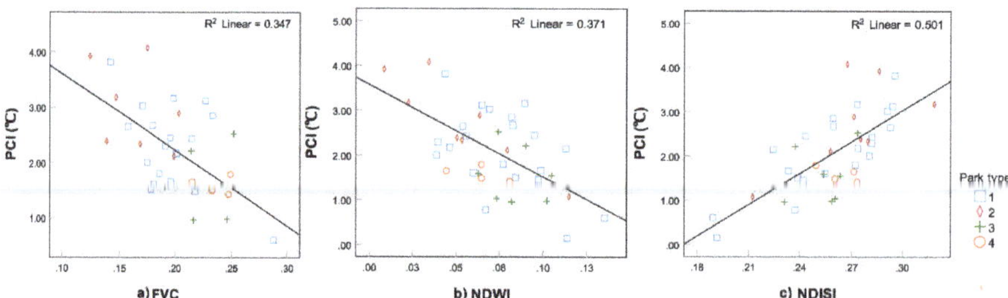

**Figure 7.** Regression analysis between the PCI and the impact factor of park surrounding area (500 m buffer). (**a**) surrounding FVC; (**b**) surrounding NDWI; (**c**) surrounding NDISI, park type: LST range (29–30 °C) based selection of 43 parks: 1-urban park; 2-theme park; 3-street park; 4-linear park.

The linear regression analysis was used to analyze the PCI relationship with the three factors outside the parks. The results show (Figure 7) that in the case of parks within the LST range of 29–30 °C the type of land cover around the park has a significant impact on PCI. PCI has a negative correlation with surrounding vegetation and water bodies, and a positive correlation with impervious surfaces in cases we analyzed from elements within the same LST range (29–30 °C). This shows that PCI is not only affected by the internal factors of the park but also related to the surrounding environment.

In addition, we analyzed the location factor on PCI based on the city rings. Zhengzhou city has three rings defined by the urban ring road (Figure 2b), the first ring is the urban center area, which is denser than the other two. The parks in the first ring have the highest average PCI (Figure 8a), as the land cover types in the urban center are mostly commercial areas and built-up areas with tall buildings and impervious surfaces, which are warmer than other areas of the city. The third ring is the low-density urban area and is covered with more green spaces and mostly low-rise buildings. We recognized that PCI is also influenced by the location factor, which is partly in relation to the different land cover types of park surroundings.

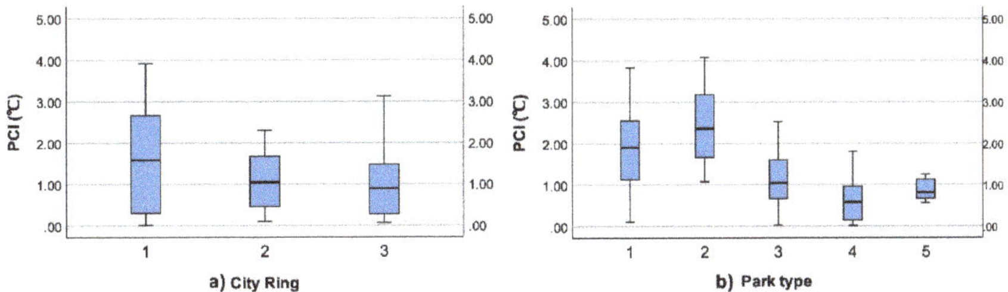


**Figure 8.** Analysis between PCI and location factor and park types (all 123 samples). (**a**) PCI and park location in three city rings; (**b**) PCI and five park types: 1-urban park; 2-theme park; 3-street park; 4-linear park; 5-urban square.

The park type can be defined based on different surrounding types, for example, linear park is mostly located, and surrounded by road or river, the urban square is usually located in the high-density area. Due to this reason, the linear park and urban square show low PCI (Figure 8b). These results are mainly attributed to the different surrounding environments and land cover of different park types. So, we can conclude that PCI is also related to the surrounding land cover types.

## 5. Discussion

### 5.1. Impact Factors of PCI

The results of Section 4.1 show that park types can have a different impact on PCI. Among the five park types, the theme park category has the highest PCI, which reaches 2.76 °C. The reason is that the theme park has higher vegetation cover and higher water surface coverage than other types. For instance, Zhengzhou Botanical Garden, where the mean FVC, NDWI, NDISI is 0.40, 0.16, 0.19, respectively. More specifically, the vegetation coverage is higher than 50%, and the diversity of species is high as well, as its primary function is science education for citizens. The linear park category has the weakest PCI. This may be because the linear parks (Figure 2c, 4-linear park) are mostly riverside green spaces, or very close to the water surface (e.g., Riverside Park). In case of linear parks in Zhengzhou, there is small LST difference between the park and its neighboring water surfaces. Therefore, the linear park type's average PCI of is the lowest.

For the results of PCI and its impact factors, we have similar conclusions. The FVC, NDWI, NDISI regression coefficient of determination ($R^2$) are 0.237, 0.433, 0.618, respectively. This means the complex correlation between PCI and park characteristics cannot be represented only by those three factors. The park patch metrics (size, fractal dimension, perimeter-area ratio, and shape index) also could not determine alone the PCI variance. As we can recognize from Section 4.3 PCI is also related to the types of the surrounding areas (Figures 6–8). High FVC and NDWI in surrounding areas make the buffer LST closer to the park internal LST, which results in low PCI. There is a positive effect between PCI and surrounding NDISI, related to the surrounding land cover types. As the result shown in Figure 8, the location factor and park type factor also effect PCI.

The cooling effect of the park can be explained from the perspective of thermal balance [39]. We can use the heat transfer theory (Bowen ratio) as an analogy to explain some of the results of this article. The Bowen ratio is the ratio of sensible heat flux to latent heat flux [60]. The surrounding areas are heat sources because the heat capacity of these is significantly smaller than the heat capacity of the parks. In heat conduction, the thermal power (sensible heat flux) absorbed by the parks from the surroundings should be equal to the excess energy resistance by photosynthesis and transpiration (latent heat flux), thus the heat conduction reaches balance. A larger green space means more energy is dissipated which results in more conducted thermal energy. Therefore, parks with large sizes, high vegetation coverage, and high water surface rate have greater energy resistance, which reduces Bowen ratio, and finally, result in higher PCI.

Furthermore, the heat conduction can also explain why parks with high Paratio and fractal dimension have lower PCI. High Paratio and fractal dimension mean that the park boundary is in a large contact surface (complex edges) with the surrounding heat sources, which is conducive to heat conduction and heat exchange. This causes temperature difference decreases, resulting in lower PCI. At the same time, this can also explain the relationship between PCI and surrounding land cover. The ambient temperature also affects the heat transfer. As a whole, to increase the cooling effect of the park, it is recommended to consider the factors of the park itself, improve the resistance to the thermal environment, and increase latent heating, so as to reduce the heat island.

*5.2. Impact Factors of Park LST*

The results of Section 4.2 reveal that high FVC, high NDWI will contribute to low park LST. Those findings are consistent with the results of the previous studies at the city level [57,59,61]. This is because the high rate of vegetation cover stores less solar energy and thus solar heat gain. The plants photosynthesis and transpiration absorb the heat during those processes [5]. Those altogether lead to lowering the park LST. In remote sensing technology, NDWI mostly represents the water body and the vegetation surface. This result also coincides with the findings in another study [62]. A recommendation for planning purposes would be to increase the vegetation and water body ratio to decrease the park LST. As NDISI had been used successfully in previous studies [57,63], we have used it to analyze the relationship of impervious surface to mean park LST. The results show that NDISI has a strong correlation with mean park LST. The reason is that the impervious surfaces have high thermal conductivity and low heat capacity [6], which lead to high LST. However, the impervious surface is an important part of park design, but we should optimize the surface rate within the design.

In terms of the results of park characteristics like the size, fractal dimension (Frac_Dim), perimeter-area ratio (Paratio), and shape index (Shape_ldx) in patch level have an impact on park LST, and those independent factors reflect the park morphology. From the results, we can recognize that large size, low Perimeter-Area ratio, and low fractal dimension decreases park LST.

Despite the practical findings in this article, we have some limitations to some extent. First, the data of satellite images have its limitation to interpret the surface thermal environment; because the temperature also relates to the microclimate factors such as wind speed and direction, humidity. The results in this paper can also be explained that park LST impact factors are the main reasons. Nevertheless, from the impact factors of park LST and PCI, the coefficient of determination ($R^2$) are not high. For instance, the FVC and NDWI regression value to mean park LST are 0.489 and 0.446, and it can only reveal that vegetation cover and water surface can explain only less 50% of the mean park LST variance. But the NDISI indicators have a significant relation to park LST, as its regression coefficient is 0.926, which means the impervious surface is the most crucial factor that brings higher LST in Zhengzhou. The analysis results of park patch metrics (size, fractal dimension, perimeter-area ratio, and shape index) and their relationship to park LST is even more complicated. Moreover, previous studies showed that even the meteorological factors (wind speed, wind direction, humidity) could influence the PCI value when we use air temperature to evaluate park cooling effect [19,64,65]. In terms of future study on the park cooling effect, we should put those aspects into consideration.

In conclusion, in terms of the UHI effect mitigation, the results on park LST are more important for planners than the results of PCI. PCI is related to the factors both inside and outside the park, but the surrounding areas are far more difficult to modify or redesign. It is clear that for planners the better option is to reduce the park LST to increase the cooling effect and mitigate the UHI effect. A future research path can focus on the analysis of parameters (e.g., vegetation types, tree coverage, height of vegetation) within the park. In landscape design, it is necessary to investigate the cooling effects of various green space design examples. Further research can deal with the vegetation cover rate analysis within a green space to optimize design from UHI point of view at local scale.

### 5.3. Implications for Urban Planning and Landscape Design

According to the results we can give a reference to urban planning and landscape design in the future. The planner and designer can follow the recommendation:

(1) In urban planning and design: increase the number of theme park types in the city, increase the park size and number in a new town/district planning;

(2) In landscape design and renewal: increase the park size, plan more vegetation and water area in parks, as well as reduce the impervious surface. At the same time, in case we follow PCI aspect of decreasing UHI we could make the park shape less complex in site design with less curving boundaries and less waving edges (based on Frac_Dim), we can consider the options of lowering the perimeter area ratio of the park by designing compact layout (Figure 9). Of course, in urban planning and design there are many other aspects to be considered, such as existing ecological corridors, road network, residential areas, wind corridors, visual preferences;

(3) Add more parks (green spaces) in the area within high impervious surface ratio, in central urban area, represented by tall buildings and impervious surfaces of commercial and built-up areas. In addition, increase the park type of high cooling effect such as theme parks and urban parks in Zhengzhou.

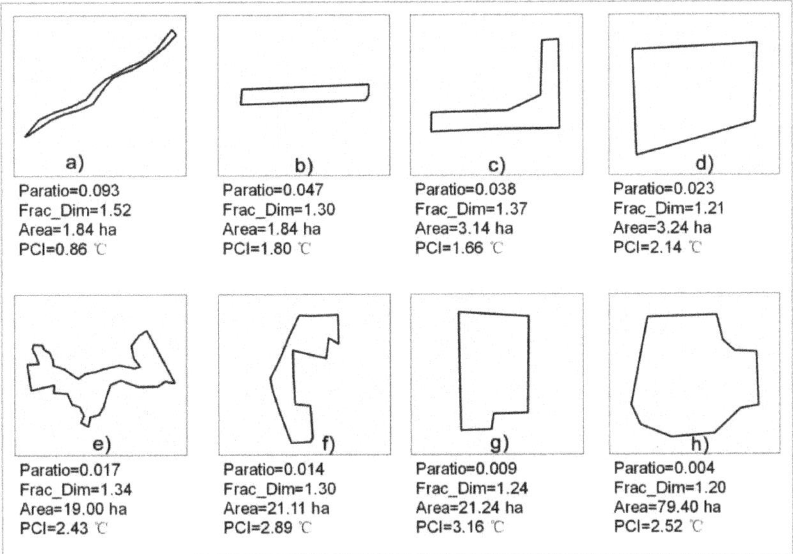


**Figure 9.** Comparison of PCI with different fractal dimensions (Frac_Dim) and perimeter area ratios (Paratio). Typical park shapes with higher Frac_Dim and Paratio values (**a,b,e,f,**) and typical shapes with lower Frac_Dim and Paratio values (**c,d,g,h**).

## 6. Conclusions

This study used a comprehensive method to investigate the urban heat island phenomenon of Zhengzhou city in China. Using the radiative transfer equation (RTE) method to retrieve the LST and we analyzed the relation of particular factors to park LST and park cooling effect. The results of this paper give a reference to characterize the complex correlation between UHI and other factors in the expanding capital city of Henan province in China. The practical results can imply urban planners and stakeholders by providing scientific guidance for future urbanization and urban management. First, the results showed that parks have a cooling effect in the city, the mean LST of the park is 0.79 °C lower than in the city. Different park types have different cooling intensity. The theme park category in Zhengzhou has the highest cooling intensity with the PCI reaching 2.76 °C. The cooling intensity of the street park and linear parks is lower; with the PCI only reaching 0. 8 °C and 0.64 °C, respectively. From the results of the park cooling effect analysis and its internal characteristics, we can recognize that the park LST depends mainly on the vegetation cover, waterbody and impervious surfaces in the park. Vegetation and water surface are the main factors of the park's cooling effect in Zhengzhou, but the impervious surface increases park LST. The study also shows a different linear correlation between the park LST and park patch metrics of Zhengzhou city. Park size and fractal dimension, perimeter area ratio affect park LST, while other geometric indicators such as shape index have no significant relation to the LST. Thus, we recommend for planners to maximize the park size and minimize perimeter area ratio to reduce the UHI effect.

The PCI is influenced by the park itself and its surrounding area:

(1)  The characteristics of the park defined by its size, perimeter area ratio and fractal dimensions all affect the PCI directly. Because of these defining factors, there seems to be a positive correlation with FVC and NDWI, while NDISI has a negative impact on the PCI;

(2)  The PCI is influenced by the surrounding land cover types coupled with the type of vegetation cover and water coverage. Because of these surrounding influences, the PCI has a directly proportional relationship to the surrounding impervious surface cover. A park's PCI has many factors to consider when understanding its mitigating effects on the UHI of its surrounding context. By first considering the factors that influence a park's temperature, we can start to change the cooling properties which help mitigate the UHI effects seen throughout cities.

We recommend additional planning consideration and construction of more parks in built-up urban areas among tall buildings and within large areas of impervious surfaces. Park types, such as theme parks and larger urban parks, should be considered with higher priority than above other park types.

Urban parks play an important role in the urban ecological environment. Targeted research on the cooling effect of parks has practical significance in urban planning and design. Parks with stronger cooling effects can help to reduce the adverse impacts of UHI problems. This study can provide a reference for urban planners and landscape architects as well as stakeholders and decision-makers.

**Author Contributions:** Data curation, H.L., G.T. and S.J.; Investigation, S.J.; Methodology, H.L.; resources, G.W.; software, H.L. and S.J.; supervision, S.J.; writing—original draft, H.L.; writing—review & editing, G.T. and S.J. All authors have read and agreed to the published version of the manuscript.

**Funding:** This research was funded by "Henan Overseas Expertise Introduction Center for Discipline Innovation" and the APC was funded by Szent István University.

**Acknowledgments:** First, we would thank the Stipendium Hungaricum Programme founding for supporting my study and research. Second, we would also thank the colleagues in the Department of Landscape Planning and Regional Development, Szent István University. Third, thanks for the 6th Fábos Conference and land journal to organize this special issue. In addition, we would also like to thank the referees and the editors for their valuable comments for improving this manuscript. Finally, I would like to especially thank Malcolm Douglas Lambson and Eleanor Jessie Potgieter from the University of the Witwatersrand (South Africa), Kelvin Andrew Lasko from Montgomery Co. Public Schools (U.S.) for helping with the English revision of this article, Thank you very much for all your efforts.

**Conflicts of Interest:** The authors declare no conflict of interest.

## References


1. Howard, L. *The Climate of London: Deduced from Meteorological Observations, Made at Different Places in the Neighbourhood of the Metropolis*; W. Phillips: Cambridge, MA, USA, 1820; Volume 1.
2. Oke, T.R. City size and the urban heat island. *Atmos. Environ.* **1973**, *7*, 769–779. [CrossRef]
3. Stewart, I.D.; Oke, T.R.; Krayenhoff, E.S. Evaluation of the 'local climate zone' scheme using temperature observations and model simulations. *Int. J. Climatol.* **2014**, *34*, 1062–1080. [CrossRef]
4. Deilami, K.; Kamruzzaman, M.D.; Liu, Y. Urban heat island effect: A systematic review of spatio-temporal factors, data, methods, and mitigation measures. *Int. J. Appl. Earth Obs. Geoinf.* **2018**, *67*, 30–42. [CrossRef]
5. US EPA, O. Heat Island Effect. Available online: https://www.epa.gov/heat-islands (accessed on 17 October 2018).
6. Yu, Z.; Guo, X.; Zeng, Y.; Koga, M.; Vejre, H. Variations in land surface temperature and cooling efficiency of green space in rapid urbanization: The case of Fuzhou city, China. *Urban For. Urban Green.* **2018**, *29*, 113–121. [CrossRef]
7. Gray, G. *EPA's 2008 Report on the Environment*; United States Environmental Protection Agency: Washington, DC, USA, 2008.
8. Voogt, J.A.; Oke, T.R. Thermal remote sensing of urban climates. *Remote Sens. Environ.* **2003**, *86*, 370–384. [CrossRef]
9. Oke, T.R. *Boundary Layer Climates*, 1st ed.; Taylor and Francis Group: New York, NY, USA, 1987.
10. Yuan, F.; Bauer, M.E. Comparison of impervious surface area and normalized difference vegetation index as indicators of surface urban heat island effects in Landsat imagery. *Remote Sens. Environ.* **2007**, *106*, 375–386. [CrossRef]
11. Li, J.; Song, C.; Cao, L.; Zhu, F.; Meng, X.; Wu, J. Impacts of landscape structure on surface urban heat islands: A case study of Shanghai, China. *Remote Sens. Environ.* **2011**, *115*, 3249–3263. [CrossRef]
12. Mohamed, A.A.; Odindi, J.; Mutanga, O. Land surface temperature and emissivity estimation for Urban Heat Island assessment using medium- and low-resolution space-borne sensors: A review. *Geocarto Int.* **2017**, *32*, 455–470. [CrossRef]
13. Jiang, J.; Tian, G. Analysis of the impact of Land use/Land cover change on Land Surface Temperature with Remote Sensing. *Procedia Environ. Sci.* **2010**, *2*, 571–575. [CrossRef]
14. Zhao, H.; Ren, Z.; Tan, J. The Spatial Patterns of Land Surface Temperature and Its Impact Factors: Spatial Non-Stationarity and Scale Effects Based on a Geographically-Weighted Regression Model. *Sustainability* **2018**, *10*, 2242. [CrossRef]
15. Kawashima, S.; Ishida, T.; Minomura, M.; Miwa, T. Relations between Surface Temperature and Air Temperature on a Local Scale during Winter Nights. *J. Appl. Meteorol.* **2000**, *39*, 1570–1579. [CrossRef]
16. Doick, K.J.; Peace, A.; Hutchings, T.R. The role of one large greenspace in mitigating London's nocturnal urban heat island. *Sci. Total Environ.* **2014**, *493*, 662–671. [CrossRef] [PubMed]
17. Yan, H.; Wu, F.; Dong, L. Influence of a large urban park on the local urban thermal environment. *Sci. Total Environ.* **2018**, *622–623*, 882–891. [CrossRef] [PubMed]
18. Yu, Z.; Xu, S.; Zhang, Y.; Jørgensen, G.; Vejre, H. Strong contributions of local background climate to the cooling effect of urban green vegetation. *Sci. Rep.* **2018**, *8*, 6798. [CrossRef] [PubMed]
19. Lindén, J.; Fonti, P.; Esper, J. Temporal variations in microclimate cooling induced by urban trees in Mainz, Germany. *Urban For. Urban Green.* **2016**, *20*, 198–209. [CrossRef]
20. Yahia, M.W.; Johansson, E.; Thorsson, S.; Lindberg, F.; Rasmussen, M.I. Effect of urban design on microclimate and thermal comfort outdoors in warm-humid Dar es Salaam. *Tanzan. Int. J. Biometeorol.* **2018**, *62*, 373–385. [CrossRef] [PubMed]
21. Chatterjee, S.; Khan, A.; Dinda, A.; Mithun, S.; Khatun, R.; Akbari, H.; Kusaka, H.; Mitra, C.; Bhatti, S.S.; Doan, Q.V.; et al. Simulating micro-scale thermal interactions in different building environments for mitigating urban heat islands. *Sci. Total Environ.* **2019**, *663*, 610–631. [CrossRef]
22. US EPA, O. Learn About Heat Islands. Available online: https://www.epa.gov/heat-islands/learn-about-heat-islands (accessed on 17 October 2018).

23. Ng, E.; Chen, L.; Wang, Y.; Yuan, C. A study on the cooling effects of greening in a high-density city: An experience from Hong Kong. *Build. Environ.* **2012**, *47*, 256–271. [CrossRef]

24. Rasul, A.; Balzter, H.; Smith, C.; Remedios, J.; Adamu, B.; Sobrino, J.; Srivanit, M.; Weng, Q. A Review on Remote Sensing of Urban Heat and Cool Islands. *Land* **2017**, *6*, 38. [CrossRef]

25. Oliveira, S.; Andrade, H.; Vaz, T. The cooling effect of green spaces as a contribution to the mitigation of urban heat: A case study in Lisbon. *Build. Environ.* **2011**, *46*, 2186–2194. [CrossRef]

26. Du, H.; Cai, W.; Xu, Y.; Wang, Z.; Wang, Y.; Cai, Y. Quantifying the cool island effects of urban green spaces using remote sensing Data. *Urban For. Urban Green.* **2017**, *27*, 24–31. [CrossRef]

27. Cao, X.; Onishi, A.; Chen, J.; Imura, H. Quantifying the cool island intensity of urban parks using ASTER and IKONOS data. *Landsc. Urban Plan.* **2010**, *96*, 224–231. [CrossRef]

28. Lin, W.; Yu, T.; Chang, X.; Wu, W.; Zhang, Y. Calculating cooling extents of green parks using remote sensing: Method and test. *Landsc. Urban Plan.* **2015**, *134*, 66–75. [CrossRef]

29. Kawashima, S. Effect of vegetation on surface temperature in urban and suburban areas in winter. *Energy Build.* **1990**, *15*, 465–469. [CrossRef]

30. Weng, Q.; Lu, D.; Schubring, J. Estimation of land surface temperature–vegetation abundance relationship for urban heat island studies. *Remote Sens. Environ.* **2004**, *89*, 467–483. [CrossRef]

31. Chen, X.-L.; Zhao, H.-M.; Li, P.-X.; Yin, Z.-Y. Remote sensing image-based analysis of the relationship between urban heat island and land use/cover changes. *Remote Sens. Environ.* **2006**, *104*, 133–146. [CrossRef]

32. Gartland, L. *Heat Islands: Understanding and Mitigating Heat in Urban Areas*; Earthscan: London, UK, 2008.

33. Liu, H.; Weng, Q. Seasonal variations in the relationship between landscape pattern and land surface temperature in Indianapolis, USA. *Environ. Monit. Assess.* **2008**, *144*, 199–219. [CrossRef]

34. Li, J.; Wang, X.; Wang, X.; Ma, W.; Zhang, H. Remote sensing evaluation of urban heat island and its spatial pattern of the Shanghai metropolitan area, China. *Ecol. Complex.* **2009**, *6*, 413–420. [CrossRef]

35. Estoque, R.C.; Murayama, Y.; Myint, S.W. Effects of landscape composition and pattern on land surface temperature: An urban heat island study in the megacities of Southeast Asia. *Sci. Total Environ.* **2017**, *577*, 349–359. [CrossRef]

36. Kolokotroni, M.; Giridharan, R. Urban heat island intensity in London: An investigation of the impact of physical characteristics on changes in outdoor air temperature during summer. *Sol. Energy* **2008**, *82*, 986–998. [CrossRef]

37. Kato, S.; Hiyama, K. *Ventilating Cities: Air-Flow Criteria for Healthy and Comfortable Urban Living*; Springer Science & Business Media: Berlin, Germany, 2012.

38. Kanda, M.; Moriizumi, T. Momentum and Heat Transfer over Urban-like Surfaces. *Bound. Layer Meteorol.* **2009**, *131*, 385–401. [CrossRef]

39. Monteith, J.L.; Unsworth, M.H. *Principles of Environmental Physics: Plants, Animals, and the Atmosphere*, 4th ed.; Elsevier/Academic Press: Amsterdam, The Netherlands, 2013.

40. Kotthaus, S.; Grimmond, C.S.B. Energy exchange in a dense urban environment-Part I: Temporal variability of long-term observations in central London. *Urban Clim.* **2014**, *10*, 261–280. [CrossRef]

41. Min, M.; Zhao, H.; Miao, C. Spatio-Temporal Evolution Analysis of the Urban Heat Island: A Case Study of Zhengzhou City, China. *Sustainability* **2018**, *10*, 1992. [CrossRef]

42. Wang, Z.; Mao, P.; Yang, H.; Zhao, Y.; He, T.; Dawson, R.J.; Li, Z. Measuring the Urban Land Surface Temperature Variations in Zhengzhou City Using the Landsat-Like Data. Available online: https://www.preprints.org/manuscript/201809.0192/v1 (accessed on 17 October 2018).

43. Urban construction over the years-Zhengzhou Statistics Bureau. Available online: http://tjj.zhengzhou.gov.cn/ndsj/216688.jhtml (accessed on 17 October 2018).

44. Mu, B.; Mayer, A.L.; He, R.; Tian, G. Land use dynamics and policy implications in Central China: A case study of Zhengzhou. *Cities* **2016**, *58*, 39–49. [CrossRef]

45. Price, J.C. Estimating surface temperatures from satellite thermal infrared data—A simple formulation for the atmospheric effect. *Remote Sens. Environ.* **1983**, *13*, 353–361. [CrossRef]

46. Zhou, J.; Li, J.; Zhang, L.; Hu, D.; Zhan, W. Intercomparison of methods for estimating land surface temperature from a Landsat-5 TM image in an arid region with low water vapour in the atmosphere. *Int. J. Remote Sens.* **2012**, *33*, 2582–2602. [CrossRef]

47. Sekertekin, A. Validation of Physical Radiative Transfer Equation-Based Land Surface Temperature Using Landsat 8 Satellite Imagery and SURFRAD in-situ Measurements. *J. Atmos. Sol. Terr. Phys.* **2019**, *196*, 105161. [CrossRef]

48. Sobrino, J.A.; Jimenez-Munoz, J.C. Minimum configuration of thermal infrared bands for land surface temperature and emissivity estimation in the context of potential future missions. *Remote Sens. Environ.* **2014**, *148*, 158–167. [CrossRef]

49. Department of the Interior, U.S. Geological Survey Landsat 8 Data Users Handbook. Available online: https://www.usgs.gov/land-resources/nli/landsat/landsat-8-data-users-handbook (accessed on 17 October 2018).

50. Carlson, T.N.; Ripley, D.A. On the relation between NDVI, fractional vegetation cover, and leaf area index. *Remote Sens. Environ.* **1997**, *62*, 241–252. [CrossRef]

51. Spronken-Smith, R.A.; Oke, T.R. The thermal regime of urban parks in two cities with different summer climates. *Int. J. Remote Sens.* **1998**, *19*, 2085–2104. [CrossRef]

52. Chang, C.-R.; Li, M.-H.; Chang, S.-D. A preliminary study on the local cool-island intensity of Taipei city parks. *Landsc. Urban Plan.* **2007**, *80*, 386–395. [CrossRef]

53. Li, Y.; Zhang, H.; Kainz, W. Monitoring patterns of urban heat islands of the fast-growing Shanghai metropolis, China: Using time-series of Landsat TM/ETM+ data. *Int. J. Appl. Earth Obs. Geoinf.* **2012**, *19*, 127–138. [CrossRef]

54. Leitao, A.B.; Miller, J.; Ahern, J.; McGarigal, K. *Measuring Landscapes: A Planner's Handbook*; Island Press: Washington, DC, USA, 2012; ISBN 978-1-59726-772-4.

55. Gao, B. NDWI—A normalized difference water index for remote sensing of vegetation liquid water from space. *Remote Sens. Environ.* **1996**, *58*, 257–266. [CrossRef]

56. Kikon, N.; Singh, P.; Singh, S.K.; Vyas, A. Assessment of urban heat islands (UHI) of Noida City, India using multi-temporal satellite data. *Sustain. Cities Soc.* **2016**, *22*, 19–28. [CrossRef]

57. Xu, H. Analysis of impervious surface and its impact on urban heat environment using the normalized difference impervious surface index (NDISI). *Photogramm. Eng. Remote Sens.* **2010**, *76*, 557–565. [CrossRef]

58. Zhang, Y.; Odeh, I.O.A.; Ramadan, E. Assessment of land surface temperature in relation to landscape metrics and fractional vegetation cover in an urban/peri-urban region using Landsat data. *Int. J. Remote Sens.* **2013**, *34*, 168–189. [CrossRef]

59. Han-qiu, X. A Study on Information Extraction of Water Body with the Modified Normalized Difference Water Index (MNDWI). *J. Remote Sens.* **2005**, *9*, 589–595.

60. Bowen, I.S. The Ratio of Heat Losses by Conduction and by Evaporation from any Water Surface. *Phys. Rev.* **1926**, *27*, 779–787. [CrossRef]

61. Jiang, Z.; Huete, A.R.; Chen, J.; Chen, Y.; Li, J.; Yan, G.; Zhang, X. Analysis of NDVI and scaled difference vegetation index retrievals of vegetation fraction. *Remote Sens. Environ.* **2006**, *101*, 366–378. [CrossRef]

62. Xu, H. Modification of normalised difference water index (NDWI) to enhance open water features in remotely sensed imagery. *Int. J. Remote Sens.* **2006**, *27*, 3025–3033. [CrossRef]

63. Sun, Z.; Wang, C.; Guo, H.; Shang, R. A Modified Normalized Difference Impervious Surface Index (MNDISI) for Automatic Urban Mapping from Landsat Imagery. *Remote Sens.* **2017**, *9*, 942. [CrossRef]

64. Wang, Y.; Bakker, F.; de Groot, R.; Wörtche, H. Effects of urban green infrastructure (UGI) on local outdoor microclimate during the growing season. *Environ. Monit. Assess.* **2015**, *187*, 732. [CrossRef] [PubMed]

65. Antoniadis, D.; Katsoulas, N.; Kittas, C. Simulation of schoolyard's microclimate and human thermal comfort under Mediterranean climate conditions: Effects of trees and green structures. *Int. J. Biometeorol.* **2018**, *62*, 2025–2036. [CrossRef] [PubMed]

 *land*

*Article*

# The Use of Community Greenways: A Case Study on A Linear Greenway Space in High Dense Residential Areas, Guangzhou

**Wenxiu Chi [1] and Guangsi Lin [1,2,3,]***

[1] South China University of Technology, School of Architecture, Department of Landscape Architecture, Guangzhou 510641, China; wendychi108@foxmail.com
[2] State Key Laboratory of Subtropical Building Science, Guangzhou 510641, China
[3] Guangzhou Municipal Key Laboratory of Landscape Architecture, Guangzhou 510641, China
* Correspondence: asilin@126.com

Received: 31 October 2019; Accepted: 6 December 2019; Published: 8 December 2019

**Abstract:** The community greenway is a kind of greenway that goes through high-density residential areas in the city and is closely related to residents' life. However, few scholars focus on how this type of greenways serves the everyday life of the community as an integrated resource. This aspect is important because the everyday life in the public space involves multiple activities. How to coordinate and satisfy these activities relates to the benefits of community greenways. Therefore, this paper takes a representative community greenway in Haizhu District of Guangzhou as an example, to study whether community greenways match the needs of necessary activities, optional activities and social activities. The usage patterns, the evaluation of the current status, the impact on everyday activities, and the importance of different construction factors were surveyed. The applied methods include site observation, questionnaires and interviews. The results show that more than 90% of users are from communities within 1 mile from the community greenway. More than half of the users (55%) are satisfied with the community greenways. Furthermore, the community greenways benefit the everyday activities of residents, such as transportation, recreation, social interaction and also other minor but important everyday activities. However, from the perspective of residents' requirements for construction factors, the status of service facilities needs to be improved. The characteristics, overall benefits, and construction implications of community greenways are therefore discussed. Community greenways can be important open space for residents and this paper is significant on community greenways meeting the needs of residents' everyday activities, thus, to provide a better community living environment and to build a better urban open space system.

**Keywords:** greenway; community greenway; everyday activities; use patterns; resident evaluation; high density residential areas; everyday public space; living environment

## 1. Introduction

The greenway movement has been recognized worldwide since its inception in the 1990s. The Greenway, which can be considered as "a linear open space established along either a natural corridor, such as a riverfront, stream valley, or ridgeline, or overland along railroad right-of-way converted to recreational use, a canal, a scenic road, or other route" [1], has unique advantages as open space when the land becomes less and less in city areas. It is a part of urban infrastructure and forms connected networks that support both ecological and social activities and process [2]. Although the urban greenways have developed into various forms according to different physical environments, its essence is to connect the residential area and neighborhoods with open space through a network-like linear landscape corridor [3–5], so that the leisure space is expanded in urban areas. In summary, urban greenways can not only adapt to the situation to the shortage of urban construction land, but also meet the increasing demand of outdoor leisure, adapting to the expansion of urbanization [6]. However, as the city areas become more and more densified, there has been a lack of space for new urban green space [7,8], including greenways. Issues have arisen, for example, many greenways cannot match the needs of the users [9]. On one hand, some greenways in built areas are based on the sidewalks and are too narrow to support the outdoor exercises [10]. On the other hand, to improve the ecological value of greenways, many of them are built outside the city [11], increasing the distance between city users and green spaces. The development of the society has led to even higher demands for open spaces [12]. Thus for urban greenways, a big challenge is that how to better serve residents instead of being a waste public sources.

Urban greenways usually have multi-functions [13], such as ecological services, recreation, commuting, and economic development [5]. To meet the challenge, the first step is to get a deep understanding of the greenway functions. One of the most important functions of urban greenways is to improve biodiversity [5]. More native species would inhabit greenways especially the urban stream corridors when the environment is renovated [14]. For recreational use, urban greenways are built to serve the urban residents by improving the natural and physical environment [15], helping people to get more outdoor activities such as physical exercise, recreation and enjoying the scenery [16–19], thus improving their physical and mental health [17]. A study indicates that people who use the greenways to exercise are more likely to meet national health standards [20]. The health benefits of greenways are impacted by the distance between greenways and residential areas, which especially increase first and then decrease in a certain distance [21]. The social benefit is also one of the important influences of greenways. On one hand, residents seek to improve community relationships in greenways [22] as the greenways near residential areas attract people to go outdoors, which can increase the frequency of neighbors' meeting [22,23]. Meanwhile, the greenways can be used as a place for community residents and their families to have fun and communicate. Through participating in these collective activities, the neighborhood interaction and community cohesion would be enhanced [15,23,24]. On the other hand, greenways help to improve social equity by serving users of a diverse sociodemographic background, especially in developing countries [9]. Urban greenways also bear important traffic functions. Many urban greenways are the essential roads for people to go to work and do shopping, which also connect communities, parks and important urban service facilities. In this way, people can reach their destinations such as shops, restaurants and transportation stations through greenways in a short time, which will improve people's travel efficiency [25–27]. In addition to the above functions, other functions of urban greenways have been further expanded recently. Some urban greenways can be used as places for urban commercial activities, allowing small vendors to enter the greenway space through time-sharing. This phenomenon is especially concentrated in the greenways close to residents or within the residential areas [28]. What's more, urban greenways also have certain impacts on the economy of residential areas. For example, urban greenways can improve the property value of nearby owners [23]. Of all the above functions, recreation is the one that the users care about the most, and satisfaction for greenways would decline if the economy function is too strong [22].

In previous studies, scholars often regarded urban greenways as a whole, while the functions of a certain greenway may be affected by its location. Greenways can show different characteristics according to different environments [5]. For example, the waterfront greenway owns better natural environment, while the greenway next to the city road has more convenient transportation. The biodiversity and recreation functions of different urban greenways will affect people's usage modes [14], which in turn affect the main functions of greenways. Among all the influencing factors, accessibility and the distance between greenways and residential areas are the most important ones [29,30]. This is because whether the greenway is easy to reach or leave directly affects the frequency of people visiting the greenway [5,31]. Greenways close to residential areas are accessed more frequently by nearby residents [15,29], and can perform various functions [23]. Through literature research, the characteristics of such greenways can be summarized: they are next to urban communities or go through residential areas, and mainly serve nearby communities [13,15,29,32], so they can be defined as "community greenways". There have been plenty cases of community greenways, such as Hudson River waterfront Greenway in New York and Kameido Ryokudo Park in Japan (Figure 1). In Nanshan District of Shenzhen, China, the community greenway density reaches 1.08 km/km$^2$ [33] because there is requirement for constructing greenways near residential area in Pearl River Delta. Though community greenways have been common in many places, there is not yet an academic consensus on community greenways.

**Figure 1.** Mappings of the cases of community greenways ((**a**) Hudson River waterfront Greenway in New York; (**b**) Kameido Ryokudo Park in Japan) (Source: the authors redrawing of the base map in mapbox).

In the authors' opinion, the community greenway can be regarded as branch "urban greenways" or "local trails" by the classification of greenways in the previous literature. Urban greenways include higher levels of development and have high levels of access to densely populated areas [5] and local trails were trails where more than 50% of the respondents are from a distance of 5 miles or less [29]. Community greenways can be subdivided from the above two categories because community greenways are even closer to residential areas and serve a smaller area. This could lead to different characteristics in the daily interactions between residents and community greenways. Some studies have focused on community greenways. For example, Akpinar studied the use patterns and factors influencing the use of Kosuyolu Urban Greenway (KUG) in Turkey, which is located within 1000 m for 30,000 inhabitants [13] and Wang did research on the construction of Furong greenway in Shenzhen, which is beside seven communities [34]. Community greenways have the main functions of urban greenways: providing leisure space and alternative transportation paths and connecting communities with nearby parks and service facilities. The main characteristics of community greenways is the location, which is close to or deep inside communities and extends to people's everyday living space. The interaction between residents and community greenways may not only affect aspects of recreation and exercise, but also in daily transportation, communication and even in the everyday activities in front of the house. A previous study has shown that greenways can improve the quality of lives [15]. Nevertheless, there is still a lack of research that focus on the activities happening in community greenways and the integrated benefits of such space. So, the question is: how do community greenways serve for the everyday activities of residents as an overall resource instead of just a recreational resource? Scholars have regarded greenways in communities as leisure space, while overlooked the multi-functions of greenways as an integrated resource to the community [5]. This leads to the isolation of greenways from the everyday lives of the residents, thus ignoring the potential benefits of the greenways.

This study takes community greenways as a specific type of urban greenways and analyzes how community greenways serve the residents from the perspective of everyday life in public space. This life is continuous in time and also public in space [35]. Applying ideas in "life between buildings", the activities in communities can be divided into necessary activities, optional activities and social activities [36], including all public activities in open space. Recreation is only a part of the optional activities. As the public space "beside the house", community greenways shorten the distance between public space and residents' life. From the perspective of everyday life, a vibrant public space can performs the pre-designed function, but also allows residents to creatively expand the functions of the space [37]. The public space should be related to the lives of the residents. Therefore, this study focuses on the relationship of community greenways and residents' daily life. The three specific questions include: (1) How do residents use community greenways in their daily life? (2) How do residents evaluate community greenways in everyday activities? (3) What are the key elements that residents consider in the construction of community greenways? From the above three aspects, the authors study on the questions how community greenways serve for residents' daily life and seek to build a better community environment.

## 2. Materials and Methods

### 2.1. Site Selection

The following principles were referenced when selecting the study site: (1) The community greenways should be part of the urban greenway networks instead of isolated trails; (2) the community greenways should be constructed in recent years to ensure that the greenways are in normal use; (3) the surrounding communities and residential groups should be diverse, ensuring that users cover a wide range of people.

Based on the above three principles, the authors conducted a pre-study on a number of community greenways in Guangzhou, China. In the end, a community greenway along Ma River, Haizhu District, Guangzhou was selected as the research site. Haizhu District is the downtown of Guangzhou, located in the south of the Pearl River. Since 2010, Haizhu District has built a relatively complete greenway network with 170 km long [38] and is a representative area of Guangzhou Greenway. From the overlay of the Greenway Network and the residential area map, the community greenway in Haizhu district is about 34.5 km long. The density of community greenway is 0.38 km/km$^2$. As most of the community greenways are located in the west of Haizhu District, the density in the west is 0.82 km/km$^2$.

The community greenway was selected due to its connection with the Guangzhou Urban Greenway Network, its relatively recent construction (the last construction was in 2017), and its location in dense communities (Figure 2). As Figure 3 illustrates, the selected area is between two urban avenue, about 1.5 miles with a width of 1.5–5 m. Located along Ma River, the Pearl River tributary, the original vegetation is preserved well and the natural environment is in good condition. On the south bank, the residential areas are mainly crowded communities which were built on the basis of old villages, namely 'urban village' in China. There are five entrances for the communities. On the north bank, the communities are mainly newly built ones. The greenway is separated from the communities by walls and there are six entrances along the greenway. A parking lot was built on the north bank, serving the nearby community and with a free classical garden, the functions of the community greenway is enriched. Besides there is diverse land use around the community greenway. The greenway connects city parks, shopping malls, transportation sites and other service facilities. The community greenway was selected mainly for the reason that the nearby areas are densely populated, about 19,300 people per square mile, which makes the greenway one of the liveliest linear open space areas in the high density urban center.

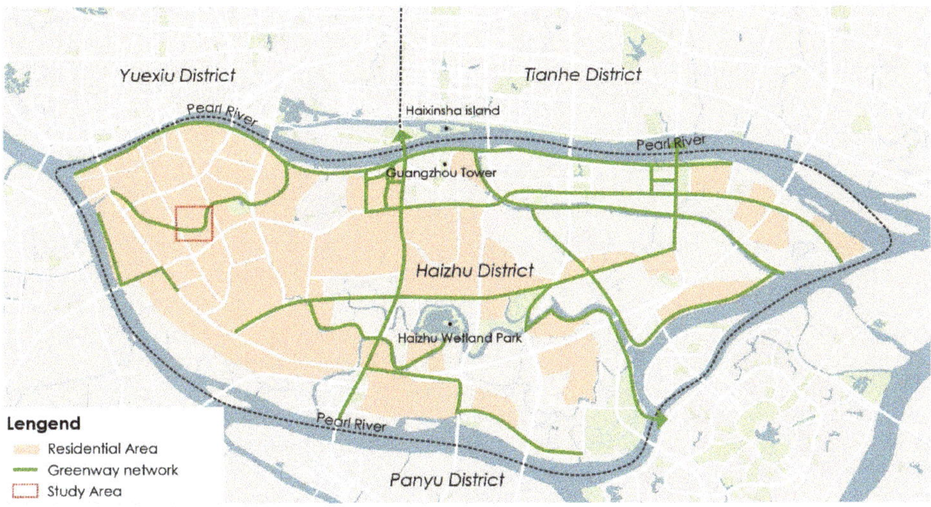



**Figure 2.** The greenway network of Haizhu District, Guanghzou (Source: the author's self-drawing).

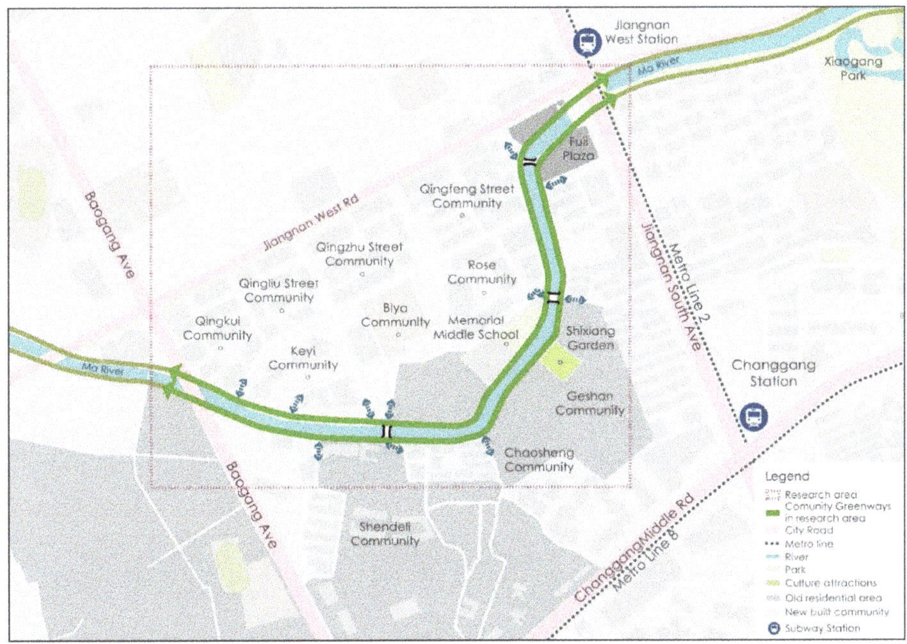



**Figure 3.** Maps of the Changgang Community Greenway, part of the Greenway in Haizhu District, Guangzhou (Source: the author's self-drawing).

## 2.2. Data Collection

The study started with site observation, guided by POE (Post Occupancy Evaluation) method [39]. The authors aimed to record the diverse activities on site and the relationship between activities and spaces. After that, questionnaires and structured interviews were conducted several times in January, September and October 2019, covering the daytime of both workdays and weekends by using intercept surveys. The intercept survey has been proved to be practical in previous studies on greenway research [32,40]. Two questionnaires were used during the survey. The first one was to study site demographics and usage patterns as adapted from Gehl's theory [36]. According to the analysis and summary of the results of this questionnaire, the second questionnaire was designed for the evaluation of the community greenway. We collected data along the community greenway on October 2 and 4, 2019 respectively for the two questionnaires. The users were randomly instructed to cover diverse groups of people.

For the study instrument, the two questionnaires were used to study the demographic information, use patterns, evaluation and opinions on different construction elements. For use patterns, residents were asked how they came to the community greenway, how long it took by walking, how often they used it, how long they usually stayed and what kind of activities they did (commuting/recreation/exercising/shopping/interacting with neighbors and family members). The evaluation included the conditions of construction elements (accessibility, traffic environment, facilities, activity space, connectivity with other urban living facilities, and overall satisfaction with the community greenway) and the impact of community greenway on the daily life (transportation, leisure activity, neighborhood activity, family activity and shopping). The options provided are 'very satisfied', 'satisfied', 'general', 'dissatisfied', 'very dissatisfied'. The items are based on previous research on greenways [13,15,24,32,41,42] and the results of observations. To learn residents' evaluation on the importance of different construction elements, participants were asked to score the importance of factors in construction (very important/important/general/unimportant/very unimportant). Besides, the results were combined with structural interviews, recording other concerning factors.

*2.3. Data Analysis*

All data was analyzed with SPSS 23.0. Multivariate regression analysis was used to investigate the relationship between demographic information, time to reach the greenway, transportation and frequency of community greenway use, and also the relationship between the evaluation of status of the greenway construction and overall satisfaction, and the impact of greenways on different activities of residents' respectively. Cross-analysis was used to study the preferences of everyday activities of different groups of community residents. In terms of residents' evaluation of the greenway and the opinions of construction factors, first, the residents' evaluation of the current status of the community greenway and the impact level of the greenway on everyday activities were analyzed through descriptive analysis. In addition, the descriptive analysis was used to study on how residents are concerned with construction elements, through the mean and the proportion of "very important" and "important". The results were presented through unstandardized coefficients, SE and 95% confidence intervals (CI). P-values of 0.05, 0.01 and 0.001 were used to indicate statistical significance. Descriptive analysis was performed with reliability analysis to ensure the credibility of the results of the questionnaire.

## 3. Result

*3.1. Site Observation*

The ped and bike system is a path shared by non-motor vehicles and pedestrian, about 1.5–7 m wide (Figure 4a). There were about 20 cyclists passing in 10 min during the peak time in the afternoon, including 8 deliverymen passing with fast speed. No extra space for recreation is left for pedestrians in the narrow section where there was only 2 m width. There is no clear bicycle parking area along the greenway. Two roads cross the survey section. There are a few stone seats along the river. These seats are lack maintenance (Figure 4b). During the peak period, the seats can be used by a percentage of 100%. The greening system can be divided into two parts. The most important vegetation is the original trees kept on the banks of the river, which are lush (Figure 4c). In addition, some of the surrounding community walls are separated from the greenway by 1–2 m wide vegetation. There are very few streetlights. Neither physical exercise facility nor public toilet was found.

(a)          (b)          (c)

**Figure 4.** The depiction of the community greenway ((**a**) the bike and ped path; (**b**) a stone seat; (c) lush vegetation) (Source: Author's own photographs).

Through observation, it was found that the common activities carried out by residents were in accordance with 'life between buildings', which can be summarized as (1) necessary activities: commuting, walking dogs; (2) optional activities: Relaxing, reading, enjoying scenery, fishing, physical exercise and shopping at the vendors along the greenway; (3) Social activities: interacting with neighbors and family members (Figure 5).

| | | |
|---|---|---|
| | Shopping | |
| | Exercise | |
| | Fishing | |
| | Enjoying Sceneries | |
| Walking the dog | Reading | interact with neighbors |
| Commuting | Relaxing | interact with families |
| **Necessary activity** | **Optional activity** | **Social activity** |

**Figure 5.** The activities in the community greenway (Source: The author's own drawing).

Although the space of the community greenway is limited, it can be summarized into several different types (Figure 6). The first type is the greenway space adjacent to the outer wall of the community. The greenway is 2 m wide. There is no extra space. Most people just passed by or took a walk. A few people would do physical exercise or fishing along the river. Some people chatted with neighbors by the river. The second type is the greenway crossing a small square, with a width of 7 m. People tended to do social activities here, such as being with children and chatting with neighbors. It is also a good place to sit alone, read books, or enjoy the scenery. The third type is the green space close to the open residential buildings, that is, the entrance and exit of the residential building directly facing the greenway. The greenway is 5 m wide. Most people would not stay here and they just passed by, and some sat on the bench by the river to chat with the neighbors. The fourth type is the greenway with a small garden. The vegetation in the greenway and in the garden is relatively closed, so the space is quiet. Residents liked to sit with family members or let the children play here. The fifth type is the greenway combined with the parking lot. The greenway is 1.5 m wide. However, some parking space is used as recreational space. Most people took physical exercise and walked the dog here.

In addition to the above activities, there were also some self-organized activities along the community greenway (Figure 7), such as small vendors in the corner, clothes drying on the fence, old furniture shared by neighbors, etc. These self-organized activities showed that the community greenway, in addition to being a public space, is also a space that residents like to share private life. It is because of these creative activities that blur the boundaries between space and private life and make the space more livable.

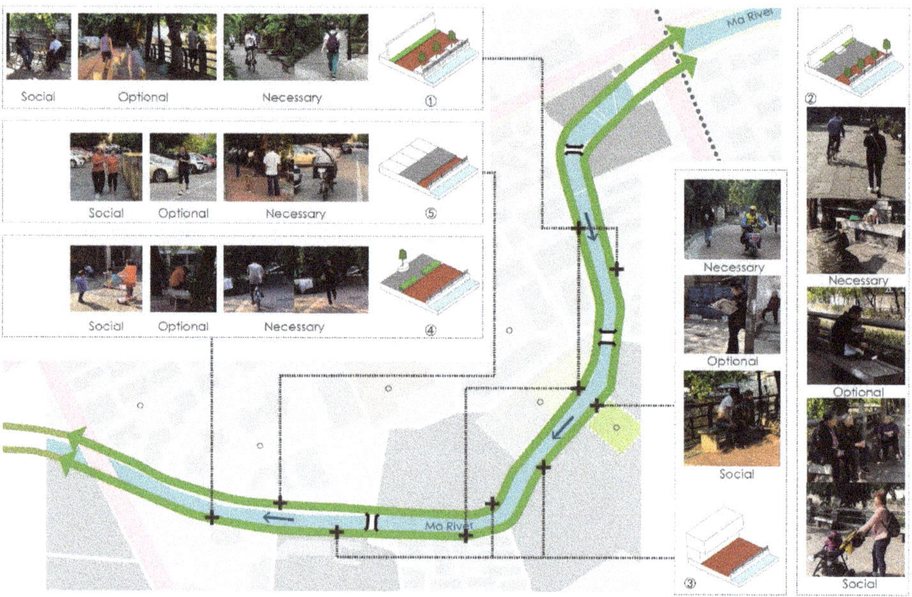

**Figure 6.** Mapping of the greenway space and public activities (Source: The author's own drawing and photographs).

(a)  (b)

**Figure 7.** The self-organized activities in the community greenways ((**a**) Haircut in the corner; (**b**) furniture shared by the neighbors) (Source: The author's own photograph).

*3.2. The Daily Use of the Community Greenway*

A total of 103 questionnaires were distributed in this round, resulting a 96% valid response rate (depending on the completion of all the questions and whether the respondents were nearby residents). We try to avoid the non-response bias through random sampling during the survey.

3.2.1. Demographics

From Table 1, the male to female residents participating in the survey was close to 1:1, which reflected to some extent that there was no significant difference in gender among the community greenway users. We observed that users of the greenway were diverse in age distribution. It should be noted that the proportion of people aged 56 and over was relatively large, accounting for 45.4%. In China, the legal retirement age for most people is between 50 and 60 years old [43,44], therefore, it can be initially determined that such people represented the retirement group. For the job status, the proportion of students was smallest (4%), and there were no significant differences of other groups. In the education level, the number of users with junior high school, high school, secondary technical school and vocational education level was 48.5%, and that of junior college and undergraduate students was 31.1%. The income level of most greenway users was 2001~10000 RMB (70.7%). According to the National Bureau of Statistics' interpretation of 2018, this interval can be understood as a medium level. From the statistics of residential streets, the vast majority of participants (91.8%) come from areas that are no more than 0.5 mile away from the community greenway, further indicating that the users of the community greenways are mainly surrounding residents.

**Table 1.** Characteristics of the study population (N = 99).

| Sociodemographic and Socioeconomic Variables | Items | No. | % of the Users |
|---|---|---|---|
| Gender | Male | 50 | 50.5 |
| | Female | 49 | 49.5 |
| Age | 18~28 years | 16 | 16.2 |
| | 29~40 years | 21 | 21.2 |
| | 41~55 years | 17 | 17.2 |
| | 56~65 years | 20 | 20.2 |
| | 66+ years | 25 | 25.2 |
| Job status | Employed | 45 | 45.4 |
| | Students | 4 | 4.0 |
| | Retired | 35 | 35.4 |
| | Unemployed | 15 | 15.2 |
| Education (highest level) | Elementary school or less | 19 | 19.2 |
| | Secondary school, High school, Secondary technical school/vocational high school | 48 | 48.5 |
| | college/Bachelor's or more | 32 | 32.3 |
| Income | RMB: <2000 monthly | 20 | 20.2 |
| | RMB:2001~5000 monthly | 43 | 43.4 |
| | RMB:5001~10000 monthly | 27 | 27.3 |
| | RMB:>10001 monthly | 9 | 9.1 |
| Area of residence | Changgang Street | 29 | 53.7 |
| | Geshan Street | 9 | 16.7 |
| | Jiangnan Street | 11 | 20.4 |
| | Other places | 5 | 9.2 |
| | Totals | 99 | 100 |

### 3.2.2. Use Patterns

From Table 2, walking was the most common way to come to the community greenway for everyday activities (84.9%). It was observed that quite a few people ride bicycles, but these people usually would not stay. Overall, 74.7% of the users lived within a 10-min walk, and 11.2% of the users lived in communities more than 20 min' walk from the greenway. 59.6% of users reported that they came to the community greenway at least once a day for leisure or other activities. About one-fifth of the users would not stay in the community greenway, and 78.8% of the users stayed in the community greenway for a period of time. In terms of user activities, 40.4% of users used community greenways for necessary activities, mainly as a traffic road. At the same time, most users (91%) used community greenways for optional activities such as leisure activities (36.4%) and physical exercise (54.6%), and some residents (8.1%) chose to shop at small vendors in the community greenway. In addition, in social activities, users interacted with family and neighborhoods.

**Table 2.** Use patterns of the community greenway.

| Categories | Items | No. | % of the Users |
|---|---|---|---|
| Transportation | Walking | 84 | 84.9 |
| | Bicycling | 10 | 10.1 |
| | Public transportation | 5 | 5.0 |
| Time to reach | <5 min | 44 | 44.4 |
| | 5–10 min | 30 | 30.3 |
| | 10–20 min | 14 | 14.1 |
| | >20 min | 11 | 11.2 |
| Frequency of use | >= 7 times per week | 59 | 59.6 |
| | 3–6 times per week | 22 | 22.2 |
| | 1–2 times per week | 12 | 12.1 |
| | Seldom | 6 | 6.1 |
| Duration of use | Would not stay | 21 | 21.2 |
| | <30 min | 31 | 31.3 |
| | 0.5–1 h | 30 | 30.3 |
| | 1–2 h | 11 | 11.1 |
| | >2 h | 6 | 6.1 |
| Activities | Daily traffic, go to destinations such as shopping malls, parks, etc. | 40 | 40.4 |
| | Leisure activities such as rest, reading, enjoying the scenery, etc. | 36 | 36.4 |
| | Physical exercise such as walking, running, dance, cycling, etc. | 54 | 54.6 |
| | Consumption, such as haircuts, purchase of fruits and vegetables | 8 | 8.1 |
| | Be with family, neighbors, friends, such as watching children, playing chess, gatherings, etc. | 9 | 9.1 |
| | Others | 4 | 4.04 |
| **Total** | | 103 | 100 |

Multivariate regression analysis was used to study the relationship between gender, age, job status, education level, monthly income, transportation, time to reach the greenway, duration of use and the frequency of use (Table 3). According to the judgment basis of $p < 0.05$, the findings indicate a significant relationship between the frequency of use and age ($\beta = -0.383$, 95% CI is $-0.399$–$0.092$, $p = 0.002$), transportation ($\beta = -0.207$, 95% CI is $-0.675$–$0.005$, $p = 0.047$) and the time to reach the greenway ($\beta = 0.411$, 95% CI $0.205$–$0.538$, $p = 0.000$). The regression results showed a negative correlation between age and frequency of use, that the older the person, the less frequently they used the community greenway. There was a negative correlation between transportation and frequency of use. Somehow there was a positive correlation between time to reach and the frequency of use. No significant correlation was found between the above factors and the duration of use ($p = 0.68 > 0.05$).

**Table 3.** Multivariate regression analysis on the relationship between demographics, transportation, time to arrive and use frequency.

| | Use Frequency ($R^2 = 0.272$) | | | |
|---|---|---|---|---|
| | β | SE | 95% C.I. | *p* |
| Gender | 0.106 | 0.169 | −0.143–0.530 | 0.256 |
| Age | −0.383 | 0.077 | −0.399–0.092 | 0.002 ** |
| Job status | 0.116 | 0.084 | −0.082–0.251 | 0.314 |
| Education | 0.068 | 0.147 | −0.206–0.377 | 0.563 |
| Income(monthly) | −0.119 | 0.122 | −0.365–0.119 | 0.315 |
| transportation | −0.207 | 0.169 | −0.675–0.005 | 0.047 * |
| Time to reach | 0.411 | 0.084 | 0.205–0.538 | 0.000 *** |

$* p < 0.05, ** p < 0.01, *** p < 0.001.$

### 3.2.3. Activity Preferences of Different Groups of People

From the results of the cross-analysis (Table 4), we can see the diversity of activities of different groups of people in the community greenway. In terms of gender, the proportion of males who exercised in community greenways (60%) was relatively high, while the proportion of women interacting with family, neighbors and friends (14.29%) was significantly higher than that of males (4%). From the perspective of age, young people tended to use the greenway for daily traffic (62.5%) while seldom for other activities (25%). A larger proportion of middle-aged and elderly people used the community greenway for optional and social activities. The elderly who were over 66 years old in the community greenway tended to do more optional (52% leisure activities, 68% physical exercise) and social activities (12% interaction with family and neighbors). While only 24% of elderly people took the greenway as traffic paths. The activities taken by residents in different job status verified the analysis of the age groups. The students mainly used the community greenway as daily passages, and the employers used it mainly for both necessary and optional activities. Retired people mainly carried out optional activities, while those who were unemployed had the highest proportion of social activities (20%). The housewives in this group accounted for a high proportion, and the main activity was to care for kids. From the influence of education level on activity, the choice of social activities (20%) mainly came from the primary school and below, mainly unemployed female and elderly people, corresponding to the above analysis of job status. For the influence of income, the users of the community greenway are mainly medium-level (2000~10000 RMB), among which the higher proportion of people in the income level of 5001~10000 RMB (59.26%) chose sports activities, and also this group had the highest proportion to shop at the vendors along the greenway (18.52%).

### 3.3. The Evaluation of the Community Greenway

The residents' evaluation of the community greenway included three parts, which were the residents' evaluation of the current status of different factors of the community greenway, the impact level of the community greenway on the residents' daily life and the importance level of the community greenway construction factors. The total reliability of the questionnaire was 0.797, which was relatively high.

Table 4. Cross-analysis of demographics and activities (N = 99).

| Categories | Items | No. (Proportion%) (N = 99) | | | | | |
|---|---|---|---|---|---|---|---|
| | | Transportation | Leisure Activity | Physical Exercise | Shopping | Be with Family and Neighbors | Others |
| Gender | Male | 17(34) | 17(34) | 30(60) | 5(10) | 2(4) | 3(6) |
| | Female | 23(46.94) | 19(38.78) | 24(48.98) | 3(6.12) | 7(14.29) | 1(2.04) |
| Age | 18~28 years | 10(62.5) | 3(18.75) | 6(37.5) | 4(25) | 0(0.00) | 0(0.00) |
| | 29~40 years | 8(38.10) | 9(42.86) | 13(61.90) | 2(9.52) | 4(19.05) | 1(4.76) |
| | 41~55 years | 8(47.06) | 6(35.29) | 7(41.18) | 0(0.00) | 1(5.88) | 2(11.76) |
| | 56~65 years | 8(40) | 5(25) | 11(55) | 1(5) | 1(5) | 1(5) |
| | 66+ years | 6(24) | 13(52) | 17(68) | 1(4) | 3(12) | 0(0.00) |
| Job status | Employed | 20(47.62) | 15(35.71) | 20(47.62) | 5(11.90) | 4(9.52) | 4(9.52) |
| | Students | 3(75) | 0(0.00) | 1(25) | 0(0.00) | 0(0.00) | 0(0.00) |
| | Retired | 12(34.29) | 16(45.71) | 24(68.57) | 1(2.86) | 2(5.71) | 0(0.00) |
| | Unemployed | 4(26.67) | 4(26.67) | 8(53.33) | 1(6.67) | 3(20) | 0(0.00) |
| Education | Elementary school or less | 5(26.32) | 8(42.11) | 11(57.89) | 1(5.26) | 4(21.05) | 1(5.26) |
| | Secondary school, High school, Secondary technical school/vocational high school | 23(47.92) | 13(27.08) | 23(47.92) | 4(8.33) | 2(4.17) | 3(6.25) |
| | college/Bachelor's or more | 12(38.71) | 15(48.39) | 20(62.5) | 3(9.68) | 3(9.68) | 0(0.00) |
| Income | RMB: <2000 monthly | 7(35) | 4(20) | 12(60) | 1(5) | 2(10) | 0(0.00) |
| | RMB:2001~5000 monthly | 17(39.53) | 18(41.86) | 21(48.84) | 2(4.65) | 4(9.30) | 2(4.65) |
| | RMB:5001~10000 monthly | 14(51.85) | 10(37.04) | 16(59.26) | 5(18.52) | 3(11.11) | 2(7.41) |
| | RMB:>10001 monthly | 2(22.22) | 4(44.44) | 5(55.56) | 0(0.00) | 0(0.00) | 0(0.00) |

The Evaluation of the Current Community Greenway and Its Impact on the Residents' Everyday Life

From Table 5, we can see accessibility scored the highest (4.19) in the status quo evaluation, and most people expressed satisfaction with the accessibility of the greenway (88%). The most dissatisfying factors was service facilities (2.97), and 76% of residents chose 'generally' or 'dissatisfied'. "There are no benches to sit and no shelter for rain. You know, the elderly need benches" "I think there is a need for public toilets here, otherwise it is very inconvenient to come here for exercise." "The lights are too dark at night, I can't see the rubbish on the road, and I would not take my kids here." Residents were relatively satisfied with other aspects. The results were 3.83 for the connection with the urban living facilities, 3.58 for the traffic environment, 3.33 for the type and amounts of the activity space. The overall satisfaction was 3.65, and 55% of residents expressed 'satisfied' with the community greenway. From the evaluation of various factors and the overall satisfaction of community greenways (Table 6), there is a positive correlation between the type and quantity of activity space and overall satisfaction ($\beta = 0.443$, 95% C.I. $= -0.222$–$0.631$, $p = 0.000$).

**Table 5.** The current status of the community greenway and its impact on the residents' daily life by mean and the proportion of "very satisfied" and "satisfied".

| | | Mean | SD. | Proportion (%) (N = 109) | |
| --- | --- | --- | --- | --- | --- |
| | | | | Very Satisfied | Satisfied |
| The residents' evaluation of the current status of different factors of the community greenway | Accessibility | 4.19 | 0.659 | 32.11 | 55.96 |
| | Transportation environment | 3.58 | 0.785 | 10.09 | 45.87 |
| | Service facilities like benches and physical facilities | 2.97 | 0.918 | 5.5 | 22.02 |
| | The types and amounts of activity space | 3.33 | 0.817 | 7.34 | 32.11 |
| | The connectivity with other urban living facilities | 3.83 | 0.788 | 19.27 | 45.54 |
| | The overall evaluation of the community greenway | 3.65 | 0.786 | 14.68 | 40.37 |
| The impact of the community greenway on the daily life | It's more convenient for transportation | 3.78 | 0.994 | 24.77 | 42.20 |
| | I take more outdoor activities alone | 3.66 | 0.993 | 20.18 | 40.28 |
| | I interact with neighbors more | 3.26 | 1.004 | 8.26 | 36.7 |
| | I take more outdoor activities with my family members | 3.31 | 1.111 | 14.68 | 32.11 |
| | The small vendors make life easier | 2.92 | 1.211 | 11.93 | 20.18 |

**Table 6.** Multivariate regression analysis on the relationship between the evaluation of current status of the construction factors and the overall satisfaction of the community greenway.

| | The Overall Evaluation of the Community Greenway ($R^2 = 0.388$) | | | |
| --- | --- | --- | --- | --- |
| | $\beta$ | SE | 95% C.I. | $p$ |
| Accessibility | 0.128 | 0.103 | −0.051–0.357 | 0.140 |
| Transportation environment | 0.039 | 0.090 | −0.138–0.217 | 0.661 |
| Service facilities like benches and physical facilities | 0.151 | 0.089 | −0.046–0.305 | 0.146 |
| The types and amounts of activity space | 0.443 | 0.103 | 0.222–0.631 | 0.000 *** |
| The connectivity with other urban living facilities | 0.056 | 0.085 | −0.112–0.224 | 0.510 |

*** $p < 0.001$.

The results of Table 5 indicate that the community greenway has significantly improved transportation (3.78) and leisure activities (3.66), and the results show that the community greenway increases the chance of interacting with neighbors (3.26) and family (3.31). The business activities in the greenway were a minority, but they also improved the lives of community residents to a certain extent (2.92). Through multiple regression analysis (Table 7), the construction status of some aspects of the greenway were related to the improvement of the daily life. When the greenway was connected to the surrounding market, shopping malls or parks, the greenway had more obvious improvement for daily transportation ($\beta$ = 0.504, 95% C.I. = 0.411–0.862). Service facilities were associated with more neighborhood interactions ($\beta$ = 0.332, 95% CI = −0.096–0.631) and family interactions ($\beta$ = 0.391, 95% CI = 0.186–0.761), while the community greenway traffic environment had a negative correlation with family interactions ($\beta$ = −0.257, 95% CI = −0.654–0.073).

**Table 7.** Multivariate regression analysis on the relationship between the evaluation of current status of the construction factors and the impact level of the community greenway on the daily life.

| | It's More Convenient for Transportation | | I Interact with Neighbors More | | I Take more Outdoor Activities with My Family Members | |
|---|---|---|---|---|---|---|
| | $\beta$ | SE | $\beta$ | SE | $\beta$ | SE |
| Transportation environment | | | | | −0.257 * | 0.146 |
| Service facilities like benches and physical facilities | | | 0.332 ** | 0.135 | 0.391 ** | 0.145 |
| The connectivity with other urban living facilities | 0.504 *** | 0.114 | | | | |

* $p < 0.05$, ** $p < 0.01$, *** $p < 0.001$.

### 3.4. The Scoring of Importance Level of the Community Greenway Construction Factors

In the evaluation of the importance of different factors relating to everyday activities in the construction of the community greenway (Table 8), sufficient service facilities scored the highest (4.11), and 81.66% of the residents considered this aspect 'very important' and 'important'. Accessibility (3.58), separation of walking trails and bicycle trails (3.66), diverse activity spaces (3.60), and the connection of urban living facilities (3.78) were not much different. Residents generally believe that these construction factors all had an important position in the construction of community greenways.

**Table 8.** The importance level of the community greenway construction factors by mean and the proportion of "very important" and "important".

| Items | Mean | SD. | Proportion (N = 109) | |
|---|---|---|---|---|
| | | | Very Important (%) | Important (%) |
| Accessibility | 3.58 | 0.913 | 12.84 | 43.12 |
| Separation of Walking trails and bicycle trails | 3.66 | 0.976 | 20.18 | 37.61 |
| Enough service facilities | 4.11 | 0.753 | 30.28 | 51.38 |
| Various activity space | 3.60 | 0.878 | 14.68 | 42.2 |
| Connection with other urban living facilities | 3.78 | 0.862 | 19.27 | 43.12 |

In addition to scoring by scales, residents also expressed some opinions on the community greenway and the construction suggestions, including the deep discussion of the issues involved in the questionnaire, as well as the content not covered by the questionnaire.

The natural environment was an issue of concern:

"The river needs to be cleaned up, because sometimes it smells bad, and it attracts a lot of mosquitoes."

(A retired old man, 60 years, about 5 min from home to the community greenway)

"The greening is very lush, but it takes up too much space. I think some vegetation can be reduced so that there would be more space for activities."

(A staff, 48 years, about 5 min from home to the community greenway)

Although the current space for activities is limited, residents understood this condition. At the same time, they tried to expand the space through innovative use and time-sharing of the community greenway:

"I would exercise in the parking lot before 6:00 in the morning. The space is relatively spacious and there is no vehicle interference at that time"

(A retired old man, 81 years, only 3 min from home to the community greenway)

For the daily activity environment of the greenway, residents generally believe that the greenway needs a safe and comfortable environment. Many residents complained about the current situation:

"The flatbed trucks that transport goods back and forth are too noisy and threaten the safety of passers-by."

(Housewife, 53 years, 8 min from home to the community greenway)

## 4. Discussion

### 4.1. The Service of Community Greenways for Different Activities

Like other urban greenways, the community greenway serves various activities. In this study, daily traffic (40.4%), leisure activities (36.4%) and physical exercises (54.6%) are the major activities. From the classification of the activities (Figure 5), it is concluded that necessary activities and optional activities are more important in the community greenway. To further analyze, different groups of people showed specific preference for activities. For example, males prefer exercise while females choose social activities more than males. This result is contrary to the previous study, which concludes that exercising was a stronger motivation for female [32]. The reason could be that in China, a large part of females prefer group dancing as a kind of exercise while in community greenways, there is limited space for this kind of activity. The social activity is an important activity for females because most retired females in China would take care of their grandsons and granddaughters, and the community greenway is a convenient open space for childcare. Different from other groups, young people mainly take the community greenway for commuting (62.5%) rather than leisure activity (18.75%) and physical exercise (37.5%). It is likely because of the lack of recreation facilities [27]. This could be an issue to be considered when constructing community greenways.

From Table 5, the most obvious improvement of the community greenway lies in necessary activity (3.78). Optional activity and social activity rank the second and the third respectively. The ranking corresponds to the status evaluation of the community greenway (Table 7). As there is positive relationship between the connectivity with other urban living facilities and transportation convenience ($\beta$ = 0.504, 95% C.I. = 0.411–0.862) and between the service facility and neighborhood communication ($\beta$ = −0.257, 95% CI = −0.654–0.073). It reflects to some extent that the convenient transportation is due to the connectivity with other urban living facilities. While the relatively weak improvement of social interactions is because of the lack of service facilities. Public space is the carrier of social interaction [45,46], as residents have shown intentions to connect with others in open spaces [47]. From this result, the improvement of service facility could be a way to increase social connections.

### 4.2. The Service of Community Greenways: Everydayness and Public Nature

The functions of a greenway have a lot to do with its location [29,32]. As a type of urban greenways near the residential area and even deep into the community space, community greenways are closely related to the everyday life of residents. The "everydayness" aspect is reflected in that community greenways mainly serve nearby neighborhood and residents tend to travel shorter distance, resulting a high-frequency use of the space. From the result of the study, we can infer that at least 90% of the users are from the surrounding communities within 1 mile. This result is consistent with a study of Denmark which showed that 84.7% of the users lived within 1000 m of green spaces [48], and the study of Turkey showing that 79.8% of the users live within 1000 m of the greenway [13]. Compared with "local trails", the data means community greenways have a smaller serving area, which shows that for nearby residents, community greenways tend to be the best choice for outdoor recreation [29]. For the frequency of use, more than half of the participants (59.6%) use 7 or more community greenways a week, consistent to Akpinar's result include that 55.4% of the users use the greenway daily [13]. Compared to a study of greenways in Shenzhen, which results that 41.5% of the users come from a distance within 1000 m and only 32.9% of the users use the greenway daily [9], community greenways promote more frequent use and activate the urban greenway network. "Public nature" is reflected in that the community greenway carries public activity, including the necessary activities (commuting, walking dogs, etc.), optional activities (enjoying the scenery, relaxing, fishing, reading, running, cycling, walking) etc.) and social activities (be with family members or neighbors, etc.) (Figure 5). In previous studies, demographic variables like gender, age, income, education level could be constraints of the use patterns of greenways [13,24]. The results of this study showed that in community greenways, users of different groups of people did not have much difference in demography information, although specific groups chose activities with their own preferences. That is to say, in general, community residents can enjoy the community greenway equally. The reason could be that as community greenways mainly serve the surrounding residential areas, residents don't differ much in social class. Another reason is that, in developing countries, all residents could freely visit and share the public spaces [9]. That is to say, community greenways promote social equity, especially in developing countries. On the other hand, community greenways also enrich the meaning of "everydayness" and "public nature" by holding everyday activities. For example, the residents use the fence in the greenway to dry clothes, or to take the furniture out to form a space gathering with neighborhood. These private activities interpret the "everydayness" from the perspective of everyday life, while the business activities in the corner add functions for leisure space, expanding the meaning of 'public nature' of community greenways. Above are the self-organized tactics of residents, showing the creativity in the community greenway, making up for the designer's strategy for the public space [37]. This shows the community greenways are closely related to the everyday life of the residents. The community greenway should not be simply regarded as leisure space, or many potential functions would be neglected.

*4.3. The Service of Community Greenways: Improving the Quality of Everyday Life*

The results show that it is widely believed that the easy accessibility of community greenways increase use, consistent with the conclusions of previous papers on greenways near the residential areas [13,15,49]. Better accessibility attracts people to use the greenway with more frequency, thereby improving people's quality of life by providing a good natural environment and social environment. In this study, users take transportation improvement and more outdoor recreation activities as the two main impacts on life. That is because, community greenways connect neighbors with shopping malls, workplaces so that the daily traffic becomes more convenient. Health benefit brought by leisure activities, physical exercise has been proved to be the most significant contribution to the users [15], and that explains why recreation use is considered one of the important functions by residents [22]. Urban greenways can be seen as the best public spaces for daily leisure, alternative transportation, and for face-to-face communication between residents [15], and from the above analysis, we can say that community greenways play a bigger role in the residents' life.

However, there are some limits of community greenways contributing to the life of residents. The community greenways are often reconstructed on the basis of the original trails. They have limited space but have complex functions at the same time. Therefore, conflicts in activities are prone to occur. For example, the transportation conflicts reflected by residents in the interview. In previous study, residents also expressed concern about community greenways, such as noise and interference caused by people walking the dog [23]. On the other hand, the use of parking lots for residents to exercise also reflects the lack of space for activities. In addition, the lack of facilities has also become a factor limiting residents to use community greenways [18].

Balancing artificial and natural environments in a high-density built environment is a huge challenge [18]. Therefore, the construction of community greenways should take into account the relationship between different activities and meet the needs of residents' activities through the transformation of space and new innovations, so as to give full play to the contribution of community greenways to residents' lives.

*4.4. Limitations of the Study*

We acknowledge several limitations for this study. First, the study focused on a representative community greenway in Haizhu District of Guangzhou. The results should be verified in more case studies in other regions. Future research should sample community greenways in other regions and compare the results to get a better understanding of how community greenways serve residents in different social contexts. In addition, investigation time was between 9:00 am and 18:00 pm during the day, not including the morning and evening time. Although the previous research on the greenway or public space did not indicate the impact of the research time on the research results [15,32,46,50]. In the interviews, the author found that a considerable number of residents like to go to the greenway in the morning or evening, so the survey may ignore people at these times.

The study classified the service of community greenways into three aspects: respectively for necessity activities, optional activities and social activities, and studied how community greenways serve the three types of activities through self-report of residents. Whether there is any correlation among these aspects still needs more research. The future research can quantify the relationship of community greenways and residents' life by constructing the relationship model with the three aspects, and to study the weight coefficient of each aspect in the overall relationship between community greenway and the resident life.

## 5. Conclusions

Community greenways, which is next to or pass through high-density residential areas, is one kind of greenways that is most closely related to people's life. This study attempts to identify the use of community greenways. The activities in community greenways include commuting, walking the dog, shopping, exercise, fishing, enjoying sceneries, reading, relaxing, interacting with neighbors or family members, which can be divided into necessary activities, optional activities and social activities. From the results, necessary activities and optional activities are the main activities, and are also the most obvious improvements that the community greenway brings to residents' life. Community greenways have two main characteristics: everydayness and public nature. The everydayness is due to its convenience to the nearby communities and high frequency of use. The public nature means that community greenways are inclusive for different users and activities. The results show that, demographically, there is not much difference in the proportion of various groups of users. Community greenways contribute to social equity especially in developing counties as residents can equally enjoy the open space. The users are relatively satisfied with the community greenway as a space for daily transportation, recreation, social interaction and so on, while some lacks also arise from the evaluation from the users. The main lack is service facilities. The results imply that it is a reason that constrains social connections. Therefore, to give full play to the advantages of the community greenway, urban planners and landscape designers should give more attention to the above aspects. If these problems can be effectively communicated and resolved, community greenways will better serve the life of residents.

**Author Contributions:** W.C. and G.L.; Data curation, W.C.; Formal analysis, W.C.; Funding acquisition, G.L.; Investigation, W.C.; Project administration, G.L.; Software, W.C.; Writing—original draft, W.C.; Writing—review and editing, G.L.

**Funding:** This research is supported by the National Natural Science Foundation of China (Grant No. 51678242 and No. 51761135025).

**Acknowledgments:** Many thanks to the help from Jiacheng, Li who majors in mathematics, and Jing, Wu, and Tianyu, Gao, friends who give suggestions to clarify and improve the paper. Their effort was highly appreciated.

**Conflicts of Interest:** The authors declare no conflicts of interest.

## References

1. Little, C.E. *Greenways for America*; Johns Hopkins University Press: Baltimore, MD, USA, 1990; p. 1.
2. Kambites, C.; Owen, S. Renewed prospects for green infrastructure planning in the UK. *Plan. Pract. Res.* **2006**, *21*, 483–496. [CrossRef]
3. Fabos, J.G. Introduction and overview: The greenway movement, uses and potentials of greenways. *Landsc. Urban Plan.* **1995**, *33*, 1–13. [CrossRef]
4. Ahern, J. Greenways as a planning strategy. *Landsc. Urban Plan.* **1995**, *33*, 131–155. [CrossRef]

5. Shafer, C.S.; Scott, D.; Mixon, J. A Greenway Classification System: Defining the Function and Character of Greenways in Urban Areas. *J. Park Recreat. Adm.* **2000**, *18*, 88–106.

6. Searns, R.M. The evolution of greenways as an adaptive urban landscape form. *Landsc. Urban Plan.* **1995**, *33*, 65–80. [CrossRef]

7. Haaland, C.; Van Den Bosch, C.K. Challenges and strategies for urban green-space planning in cities undergoing densification: A review. *Urban For. Urban Green.* **2015**, *14*, 760–771. [CrossRef]

8. Brunner, J.; Cozens, P. Where have all the trees gone? Urban consolidation and the demise of urban vegetation: A case study from Western Australia. *Plan. Pract. Res.* **2013**, *28*, 231–255. [CrossRef]

9. Chen, Y.; Gu, W.; Liu, T.; Yuan, L.; Zeng, M. Increasing the use of urban greenways in developing countries: A case study on Wutong Greenway in Shenzhen, China. *Int. J. Environ. Res. Public Health* **2017**, *14*, 554. [CrossRef]

10. Liu, Z.; Lin, Y.; De Meulder, B.; Wang, S. Can greenways perform as a new planning strategy in the Pearl River Delta, China? *Landsc. Urban Plan.* **2019**, *187*, 81–95. [CrossRef]

11. Lai, S.H.; Zhu, J. Community greenway: A new trend of greenway practice in compact cities. *Landsc. Archit.* **2012**, *34*, 77–82.

12. Thompson, C.W. Urban open space in the 21st century. *Landsc. Urban Plan.* **2002**, *60*, 59–72. [CrossRef]

13. Akpinar, A. Factors influencing the use of urban greenways: A case study of Aydın, Turkey. *Urban For. Urban Green.* **2016**, *16*, 123–131. [CrossRef]

14. Asakawa, S.; Yoshida, K.; Yabe, K. Perceptions of urban stream corridors within the greenway system of Sapporo, Japan. *Landsc. Urban Plan.* **2004**, *68*, 167–182. [CrossRef]

15. Shafer, C.S.; Lee, B.K.; Turner, S. A tale of three greenway trails: User perceptions related to quality of life. *Landsc. Urban Plan.* **2000**, *49*, 163–178. [CrossRef]

16. West, S.T.; Shores, K.A. The impacts of building a greenway on proximate residents' physical activity. *J. Phys. Act. Health* **2011**, *8*, 1092–1097. [CrossRef] [PubMed]

17. West, S.T.; Shores, K.A. Does building a greenway promote physical activity among proximate residents? *J. Phys. Act. Health* **2015**, *12*, 52–57. [CrossRef] [PubMed]

18. Liu, K.; Siu, K.W.M.; Gong, X.Y.; Gao, Y.; Lu, D. Where do networks really work? The effects of the Shenzhen greenway network on supporting physical activities. *Landsc. Urban Plan.* **2016**, *152*, 49–58. [CrossRef]

19. Auchincloss, A.H.; Michael, Y.L.; Kuder, J.F.; Shi, J.; Khan, S.; Ballester, L.S. Changes in physical activity after building a greenway in a disadvantaged urban community: A natural experiment. *Prev. Med. Rep.* **2019**, *15*, 100941. [CrossRef]

20. Price, A.E.; Reed, J.A.; Muthukrishnan, S. Trail user demographics, physical activity behaviors, and perceptions of a newly constructed greenway trail. *J. Commun. Health* **2012**, *37*, 949–956. [CrossRef]

21. Frank, L.D.; Hong, A.; Ngo, V.D. Causal evaluation of urban greenway retrofit: A longitudinal study on physical activity and sedentary behavior. *Prev. Med.* **2019**, *123*, 109–116. [CrossRef]

22. Lin, J.L.; Gan, Q.L.; Wei, S.; Zhang, T.J. Guang Zhou Greenway Function Perception of IPA Assessment and Analysis. *Yunnan Geogr. Environ. Res.* **2012**, *3*, 14–18.

23. Corning, S.E.; Mowatt, R.A.; Charles Chancellor, H. Multiuse trails: Benefits and concerns of residents and property owners. *J. Urban Plan. Dev.* **2012**, *138*, 277–285. [CrossRef]

24. Palardy, N.P.; Boley, B.B.; Gaither, C.J. Resident support for urban greenways across diverse neighborhoods: Comparing two Atlanta BeltLine segments. *Landsc. Urban Plan.* **2018**, *180*, 223–233. [CrossRef]

25. Tan, K.W. A greenway network for Singapore. *Landsc. Urban Plan.* **2006**, *76*, 45–66. [CrossRef]

26. Larson, L.R.; Keith, S.J.; Fernandez, M.; Hallo, J.C.; Shafer, C.S.; Jennings, V. Ecosystem services and urban greenways: What's the public's perspective? *Ecosyst. Serv.* **2016**, *22*, 111–116. [CrossRef]

27. Sims-Gould, J.; Race, D.L.; Vasaya, N.; McKay, H.A. A New Urban Greenway in Vancouver, British Columbia: Adolescents' Perspectives, Experiences and Vision for the Future. *J. Transp. Health* **2019**, *15*. [CrossRef]

28. Wang, F. The Research on Guangzhou Community Greenway's Development and Usage from the Perspective of Everyday Life. Master's Thesis, South China University of Technology, Guangzhou, China, 2015.

29. Gobster, P.H. Perception and use of a metropolitan greenway system for recreation. *Landsc. Urban Plan.* **1995**, *33*, 401–413. [CrossRef]

30. Zoellner, J.; Hill, J.L.; Zynda, K.; Sample, A.D.; Yadrick, K. Environmental perceptions and objective walking trail audits inform a community-based participatory research walking intervention. *Int. J. Behav. Nutr. Phys. Act.* **2012**, *9*, 6. [CrossRef]

31. Senes, G.; Rovelli, R.; Bertoni, D.; Arata, L.; Fumagalli, N.; Toccolini, A. Factors influencing greenways use: Definition of a method for estimation in the Italian context. *J. Transp. Geogr.* **2017**, *65*, 175–187. [CrossRef]

32. Keith, S.J.; Larson, L.R.; Shafer, C.S.; Hallo, J.C.; Fernandez, M. Greenway use and preferences in diverse urban communities: Implications for trail design and management. *Landsc. Urban Plan.* **2018**, *172*, 47–59. [CrossRef]

33. Liu, J.Y. Research on POE and Optimization Strategies of the Community Greenways in Nanshan, Shenzhen. Master's Thesis, Harbin Institute of Technology, Harbin, China, 2014.

34. Wang, J.; Li, Q.W.; Peng, L.Q. The Enlightenments on the Construction of Community Greenway of Guangdong—Based on the Example of Furong Grennway in Shenzhen. *Chin. Landsc. Archit.* **2014**, *5*, 22.

35. Ma, R.J. Conception Analysis of Urban Heritage of Everydaynes. *Huazhong Archit.* **2015**, *33*, 27–31.

36. Gehl, J. *Life between Buildings: Using Public Space*; Island Press: Washington, DC, USA, 2011.

37. Chase, J.L.; Crawford, M.; Kaliski, J. *Everyday Urbanism*; The Monacelli Press: New York, NY, USA, 2008; p. 224.

38. Guangdong Greenway. Available online: http://www.gzlyyl.gov.cn/slyylj/gzld/201811/39613b6e08cd44989513b0dc178868c1.shtml (accessed on 1 December 2019).

39. Marcus, C.C.; Francis, C. *People Places: Design Guidelines for Urban Open Space*; John Wiley & Sons: New York, NY, USA, 1997; p. 369.

40. Troped, P.J.; Whitcomb, H.A.; Hutto, B.; Reed, J.A.; Hooker, S.P. Reliability of a brief intercept survey for trail use behaviors. *J. Phys. Act. Health* **2009**, *6*, 775–780. [CrossRef] [PubMed]

41. Kaźmierczak, A. The contribution of local parks to neighbourhood social ties. *Landsc. Urban Plan.* **2013**, *109*, 31–44. [CrossRef]

42. Gobster, P.H.; Westphal, L.M. The human dimensions of urban greenways: Planning for recreation and related experiences. *Landsc. Urban Plan.* **2014**, *68*, 147–165. [CrossRef]

43. State Council. *Interim Measures of the State Council on the Resettlement of Old, Weak, Sick and Disabled Cadres*; Government of China: Beijing, China, 1978.

44. State Council. *Interim Measures of the State Council on the Retirement and Resignation of Workers*; Government of China: Beijing, China, 1978.

45. Nasution, A.D.; Zahrah, W. The Space is not ours, the life of public open space in gated community in Medan, Indonesia. *Procedia Soc. Behav. Sci.* **2015**, *202*, 144–151. [CrossRef]

46. Whiting, J.W.; Larson, L.R.; Green, G.T.; Kralowec, C. Outdoor recreation motivation and site preferences across diverse racial/ethnic groups: A case study of Georgia state parks. *J. Outdoor Recreat. Tour.* **2017**, *18*, 10–21. [CrossRef]

47. Dwyer, J.F.; Barro, S.C. Outdoor recreation behaviors and preferences of urban racial/ethnic groups: An example from the Chicago area. In Proceedings of the 2000 Northeastern Recreation Research Symposium, Bolton Landing, NY, USA, 2–4 April 2000; Kyle, G., Ed.; General Technical Report NE-276. US Department of Agriculture, Forest Service, Northeastern Research Station: Newtown Square, PA, USA; pp. 159–164.

48. Schipperijn, J.; Stigsdotter, U.K.; Randrup, T.B.; Troelsen, J. Influences on the use of urban green space–A case study in Odense, Denmark. *Urban For. Urban Green.* **2010**, *9*, 25–32. [CrossRef]

49. Fan, S.; Chen, X.; Ren, H.; Shen, W.; Xiao, R.; Zhang, Q.; Wu, Z.; Su, Y. Landscape structure and network characteristics of the greenway system in Guangzhou City, South China. *Landsc. Ecol. Eng.* **2019**, *15*, 25–35. [CrossRef]

50. Weber, S.; Boley, B.B.; Palardy, N.; Gaither, C.J. The impact of urban greenways on residential concerns: Findings from the Atlanta BeltLine Trail. *Landsc. Urban Plan.* **2017**, *167*, 147–156. [CrossRef]

 *land*

*Article*

# Research-Based Design Approaches in Historic Garden Renovation

**Albert Fekete [1,*] and László Kollányi [2]**

[1] Department of Garden Art, Faculty of Landscape Architecture and Urbanism, Szent István University, 1118 Budapest, Hungary
[2] Department of Landscape and Regional Planning, Faculty of Landscape Architecture and Urbanism, Szent István University, 1118 Budapest, Hungary; kollanyi.laszlo@tajk.szie.hu
* Correspondence: fekete.albert@tajk.szie.hu

Received: 2 November 2019; Accepted: 10 December 2019; Published: 12 December 2019

**Abstract:** The renewal of historic gardens, landscapes, and sites has grown to be a current issue in Central and Eastern Europe. Based on scientific research, the Department of Garden Art of the Szent István University, Faculty of Landscape Architecture and Urbanism has been dealing with landscape renewal since 1963 on regional, settlement, and garden scales, too. More than 50 years of experience has already proved the advantage of such a research-based design approach in garden and landscape renewal processes, Landscape Architecture has developed from a very practical basis. The purpose of this paper is to show the most significant conclusions of our historic garden research of castle gardens from the Carpathian Basin, focusing on the importance of visual connections designed initially on the sites. Using case studies, the paper intends to explore how proper landscape design in historic environments is achieved. The historical value cannot be simplified or understood as the notion of "old", the heritage being represented by the all-time valuable garden features and elements, independent from their formation in time. In addition to the historical authenticity of the actual use, the social needs and sustainability are important aspects, which must be integrated into heritage protection and reclamation.

**Keywords:** landscape research; view; visual link; castle garden; garden renewal; Carpathian Basin; research and design; historic garden and landscape

---

## 1. Introduction

Rediscovering the scenic value of the landscape has Renaissance roots, and recognizing the aesthetic value of the landscape can be traced in the Renaissance landscape and garden descriptions. The Renaissance image of the world through the conscious observation of nature brings a new chapter in the relationship between man and his environment: he does not only observe the landscape but also perceives its atmosphere and puts his experience of the landscape into words and pictures. Being part of the landscape, the garden belongs to the humanistic approach to life, as a real living space, and as an embodiment of beauty and harmony. The virtual expansion of the garden limits takes place: humanists consciously link the prospect of the landscape into the garden sight, and since then, this has become a consistently applied landscape composition tool—just as in the case of Baroque gardens and landscaped gardens [1–4].

The deliberate shaping of large-scale (landscape) visual connections started already in the Baroque but was accomplished in the Carpathian Basin in the landscaped gardens of the 19th century. The "picturesque" evokes the ideals of the early Enlightenment, a symbolically perceived part of the world. In this period, conscious artistic space organisation resulted in a number of spatial compositions using less functional than picturesque artificial elements (sculptures, buildings, decorations, etc.) in order to create atmosphere or staffage. These prominent design elements used as a compositional center, which also served as signals and had a symbolic meaning, marked the focal points of the visual axes [5–9].

Regardless of the era, however, it can be stated that during modern garden history, the most important compositional goal of the creators was always the design of a garden image and that of the sight of the garden in time and space; also, the creation of ideal visual links between the garden and the landscape, that is, the view meant to fascinate.

The creation and display of landscape elements is the result of a creative and conscious spatial arrangement. These elements vary depending on whether they are natural or built elements, and from this difference, a different cultural value ensues. Thus, the visual connections ensuring their display are, in some cases, difficult to acknowledge; occasionally, they even transmit different messages to different segments of the society. That is also a reason why cautious and thorough interpretation is needed when assessing their aesthetic value: alongside the examination of the physical appearance, the analysis of the emotions and of the atmosphere evoked by the landscape is also essential [10,11].

The castle garden has thousand-fold connections to its surroundings. At the time of its formation, it was not a simple ornamental garden but featured as an indispensable part of a complex, cultural-historical, ecological and—last but not least—economic system. That is what made it possible for the garden to function in the long run and be sustainable at the same time. Thus, the subject matter of the study is a consciously planned ecological-technical system, both historic and artistic, which can only be interpreted if we embed it into its wider environment. The potential of views and visual axes were taken as features that greatly determined the image of the small area forming the wider environment of the castle park. The statement that "the aristocratic mansion-park ensembles can be considered as a model situation from which the plant and knowledge elements of the model were democratically dispersed on the palace-castle-yard-civic house-farmhouse line" also emphasises the landscape and settlement character of the castle garden, as well as its role in defining the structure [12]. Therefore, when we renew our long-standing gardens while respecting the circumstances of their formation, we need to re-interpret not only the garden itself and its internal relations but their relationship and complex system of connections with the countryside and society as well.

The restoration of the visual axes defining historical compositions is one of the most important tools for the renewal of the historic garden and landscape structure. The paper illustrates and emphasises the essence of restoring the view, and shows that the results of several-decades-long research on the subject can be applied in specific design situations and locations with various features.

## 2. Materials and Methods

The systematic research on Transylvanian castle gardens started in 2004, in cooperation with the National Centre for Historic Monument Conservation and Restoration and has been ongoing research with changing partnerships since then. Until nowadays, the work has comprised the historical research, site survey, analysis, and assessment of 100 Transylvanian castle gardens (Table 1) with the participation of more than 120 university students, many teachers, and external experts contributing significantly to the knowledge base on European garden history [7,13–16].

**Table 1.** The list of castle-gardens investigated during our research.

| Hungarian (Romanian) Name of the Settlement | Name of the Owner Family | Hungarian (Romanian) Name of the Settlement | Name of the Owner Family |
| --- | --- | --- | --- |
| 1. Abafája (Apalina, MS) | Huszár castle | 51. Maroskeresztúr (Cristuru M, MS) | Knöpfler castle |
| 2. Alsózsuk (Jucul de Jos, CJ) | Kemény castle | 52. Maroshéviz (Toplita, HR) | Urmánczy castle |
| 3. Alvinc (Vintul de Jos, AB) | Martinuzzi c | 53. Marosillye (Ilia, HD) | Bornemisza c |
| 4. Aranyosgerend (Luncani, CJ) | Kemény castle | 54. Marosnémeti (Mintia, HD) | Gyulay castle |
| 5. Árkos (Arcus, CV) | Szentkereszti c | 55. Marosugra (Ogra, MS) | Haller castle |
| 6. Árokalja (Arcalia, BN) | Bethlen castle | 56. Marosújvár (Ocna Mures, AB) | Teleki castle |
| 7. Bályok (Balc, BH) | Károlyi castle | 57. Marosvécs (Brancovenesti, MS) | Kemény castle |
| 8. Bethlen (Beclean, BN) | Bethlen castle | 58. Marosszentgyörgy (Sangeorgiu de Mures, MS) | Petki-Máriaffy castle |

**Table 1.** *Cont.*

| Hungarian (Romanian) Name of the Settlement | Name of the Owner Family | Hungarian (Romanian) Name of the Settlement | Name of the Owner Family |
|---|---|---|---|
| 9. Bethlenszentmiklós (Sanmiclaus, AB) | Bethlen castle | 59. Marosszentkirály (Sancraiu de Mures, AB) | Bánffy castle |
| 10. Bihardiószeg (Diosig, BH) | Zichy castle | 60. Mácsa (Macea, AR) | Csernovics castle |
| 11. Bonchida (Bontida, CJ) | Bánffy castle | 61. Mezőzáh (Zau de Campie, MS) | Ugron castle |
| 12. Bodola—1 (Budula, CV) | Béldy castle | 62. Mezőörményes (Urmenis, MS) | Rákóczi castle |
| 13. Bodola—2 (Budula, CV) | Mikes castle | 63. Nagyernye (Ernei, MS) | Bálintitt castle |
| 14. Bonyha (Bahnea, MS) | Bethlen castle | 64. Nagykároly (Carei, SM) | Károlyi castle |
| 15. Branyicska (Branisca, HD) | Jósika castle | 65. Nagykend (Chendu, MS) | Schell castle |
| 16. Cege—1 (Taga, CJ) | Wass Á castle | 66. Nagyteremi (Tirimia, MS) | Bethlen castle |
| 17. Cege—2 (Taga, CJ) | Wass J castle | 67. Nyárádszentbenedek (Murgesti, MS) | Toldalagi castle |
| 18. Csákigorbó (Garbau, SJ) | Haller-Jósika c | 68. Őraljaboldogfalva (Santamaria Orlea, HD) | Kendeffy castle |
| 19. Csombord (Ciumbrud, AB) | Kemény castle | 69. Piski (Simeria, HD) | Ocskay-Fáy c |
| 20. Dálnok (Dalnic, CV) | Gaál castle | 70. Pusztakamarás (Camarasu, CJ) | Kemény castle |
| 21. Drág (Dragu, SJ) | Wesselényi c | 71. Radnót (Iernut, MS) | Rákóczi castle |
| 22. Erdőszentgyörgy (Sang de Padure, MS) | Rédey castle | 72. Radnótfája (Iernuteni, MS) | Matskási castle |
| 23. Fiatfalva (Filias,, HR) | Ugron castle | 73. Sarmaság (Sarmasag, SJ) | Kemény castle |
| 24. Fugad (Ciuguzel, AB) | Bánffy castle | 74. Sáromberke (Dumbravioara, MS) | Teleki castle |
| 25. Gernyeszeg (Gornesti, MS) | Teleki castle | 75. Sárpatak (Glodeni, MS) | Teleki castle |
| 26. Görgényszentimre (Gurghiu, MS) | Rákóczi castle | 76. Sepsiköröspatak (Valea Cris, CV) | Kálnoky castle |
| 27. Gyalu (Gilau, CJ) | Barcsay castle | 77. Soborsin (Savarsin, AR) | Nádasdy-Forray c |
| 28. Gyeke (Geaca, CJ) | Béldy castle | 78. Szászfenes (Floresti, CJ) | Mikes castle |
| 29. Gyergyószárhegy (Lazarea, HR) | Lázár castle | 79. Székelyhíd (Sacueni, BH) | Stubenberg castle |
| 30. Gyulafehérvár (Alba Iulia, AB) | Episcopal castle | 80. Székelyszenterzsébet (Elisei, HR) | Kemény castle |
| 31. Hadad—1 (Hodod, SM) | Wesselényi c | 81. Szentbenedek (Manastirea, CJ) | Kornis castle |
| 32. Hadad—2 (Hodod, SM) | Degenfeld c. | 82. Szentdemeter—1 (Dumitreni, MS) | Balási castle |
| 33. Kelementelke—1 (Calimanesti, MS) | Simén castle | 83. Szentdemeter—2 (Dumitreni, MS) | Schell castle |
| 34. Kelementelke—2 (Calimanesti, MS) | Henter castle | 84. Szentgothárd (Sucutard, CJ) | Wass castle |
| 35. Kendilóna (Luna de Jos, CJ) | Teleki castle | 85. Szilágybagos (Boghis, SJ) | Bánffy castle |
| 36. Kerelőszentpál (Sanpaul, MS) | Haller castle | 86. Szilágynagyfalu—1 (Nusfalau, SJ) | Bánffy castle |
| 37. Keresd (Cris, MS) | Bethlen castle | 87. Szilágynagyfalu—2 (Nusfalau, SJ) | Bánffy castle |
| 38. Kerlés (Chirales, BN) | Bethlen castle | 88. Szilágysomlyó (Simleu Silvan, SJ) | Báthory castle |
| 39. Kisbún (Boiu, MS) | Bethlen castle | 89. Székelyudvarhely (Sambatesti—Odorheiu Secuiesc, HR) | Ugron castle |
| 40. Koltó (Coltau, MM) | Teleki castle | 90. Szurdok (Surduc, SJ) | Jósika castle |
| 41. Koronka (Corunca, MS) | Toldalagi c | 91. Tasnád (Tasnad, SJ) | Cserei castle |
| 42. Korpád (Corpadea, CJ) | Gaál castle | 92. Torockószentgyörgy (Trascau, AB) | Thoroczkay- castle |
| 43. Kővárhosszúfalu (Satulung, MM) | Teleki castle | 93. Vajdaszentivány (Voivodeni, MS) | Zichy-Horváth c |
| 44. Kraszna (Crasna, SJ) | Cserei castle | 94. Válaszút (Rascruci, CJ) | Bánffy castle |
| 45. Kutyfalva (Cuci, MS) | Degenfeld c | 95. Váralmás (Almasu, SJ) | Csáky castle |
| 46. Magyarbükkös (Bichis, MS) | Kemény castle | 96. Várfalva (Moldovenesti, CJ) | Jósika castle |
| 47. Magyarcsesztve (Cisteiu de Mures, AB) | Mikes castle | 97. Zabola (Zabala, CV) | Mikes castle |
| 48. Magyarfenes (Vlaha, CJ) | Jósika castle | 98. Zám (Zam, HD) | Nopcsa castle. |
| 49. Magyarózd (Ozd, MS) | Radák-Pekry c | 99. Zsibó—1 (Jibou, SJ) | Wesselényi castle |
| 50. Magyarpécska (Pecica, AR) | Klebelsberg c | 100. Zsibó—2 (Jibou, SJ) | Béldy castle |

## 2.1. Work Methodology

The survey methodology of Transylvanian castle gardens was based on the principle that the sites concerned must be interpreted in context with the related settlements and landscapes, as the only way to understand their historical importance and current value.

For a systematic survey of the visual links and eye-catchers of the castle gardens, we established the following theoretical framework:

- Identification of all potential gardens,
- Definition of the priority list of the gardens,
- Historical research of the gardens,
- General landscape assessment of the present conditions of the gardens and its environment,
- Survey and assessment of the nowadays area, the spatial layout, the composition, the visual links of the gardens and its surroundings.

The research methodology may be practically divided into two main parts: the garden history research and the site survey.

### 2.1.1. Garden History Research

The goal of the historical research of primary and secondary sources found (archives, library and museum materials, map and postcard collections, thematic bibliography reviews, internet sources, etc.) is to provide a clear idea of the establishment and development of the gardens. It comprises the role the sites play in the landscape and the urban character and layout, and the landscape scale relationships that served as a basis for the establishment of the castle garden and determined the character of the surrounding landscapes to a great extent. The garden history research also deals with the architectural history of the castle and the family history of the owners. Family history data proved to be especially important, as it was tightly linked to the initial creation or later transformation of parks, certain garden sections, or specific garden features. The owner families are the bearers of the intellectual and cultural substance that is essential for the spirit and identity of the place, and for the establishment of the castle gardens. Many of the gardens were shaped from the ideas of the owners, or the design was directly influenced by the owners. Owners of the castles and gardens may therefore also be considered as creators of these historic monuments. The results of the garden history research are the inventory and the cultural heritage assessment.

### 2.1.2. Site Survey

The site survey precisely records the actual conditions of each castle garden (sketches, minutes, GPS coordinates, geodetic surveys, geophysical surveys, aerial remote sensing survey, plant inventory, digital photographic inventory, etc.) as well as the valuable features, and thus, it serves as a status report or basis for comparison for conservation strategies and any future restorations. A topographic map (land registry map, etc.) provides the basis for the survey of the general conditions and valuable landscape features. Definition and classification of the survey criteria were important steps of the procedure. As a starting point, we took the criteria of historical heritage surveys in Hungary, complementing and adapting it to local circumstances as necessary. The survey form records aspects of heritage conservation, visual landscape protection, and dendrology, focusing primarily on the valuable botanical, architectural and landscape features, visual links, and spatial composition. The prepared survey form also integrates the recommendations from the English Heritage and the National Centre for Historic Monument Conservation in Hungary.

In the case of one of the most famous and influential British gardens, the Rousham House Garden, we can also bring up the eye-catchers and defined visual axes as the most significant compositional tools—as proof of their success in garden history to this day: "The many wandering walks through the gardens are full of delicious surprises, a sudden meeting with a dying gladiator, a glimpse of Apollo, or a long view of a Gothic mill, an ancient bridge or distant trees, or arrival at an unexpected seat in an alcove", says Hal Moggridge, an English landscape architect [17].

Concerning Transylvanian castle gardens and landscapes, we tried to determine those prospects, eye-catchers, and visual axes in cases of 100 locations, which through their meanings and symbolic messages play an essential role in the garden composition or landscape they are part of.

Similarly to Hungarian and other examples from the countries in Europe [18], planned visual links in Transylvania are specific mostly to landscape gardens from the 19th century, a period that was also the golden age for Transylvanian castle gardens (Figure 1).

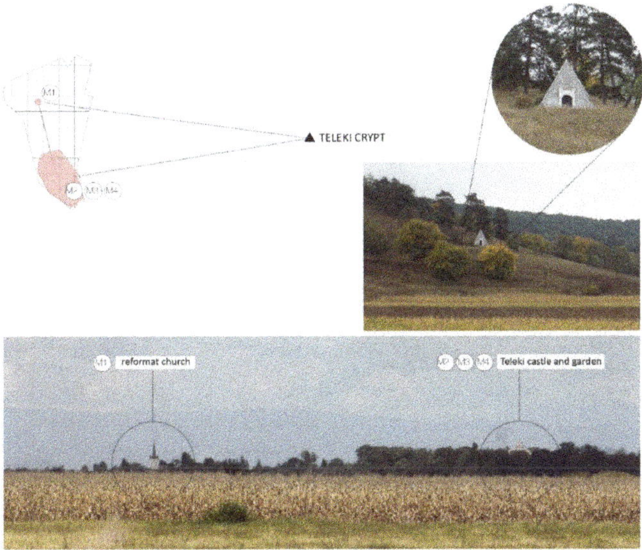


**Figure 1.** Visual attractions represented by built features in the case of the Teleki estate from Gernyeszeg (Gornesti, R.O.). Photos and graphical processing by Fekete, A., based on [19].

The deliberate spatial layout resulted in landscape compositions with structures and ornamental elements that were not really functional in themselves; instead, picturesque elements were applied as eye-catchers, staffage, and effects to create a specific atmosphere.

As a highlight of the composition, the eye-catchers also played an important role in the case of the designed landscapes in Transylvania, providing the focus for visual axes, occasionally having an additional, symbolic meaning.

## 3. Results

Besides being aware of garden history, the expert needs a thorough knowledge of the current situation, a regard for the value that still exists, as well as detailed on-site research and survey in order to deal with garden renewal.

Recognition and interpretation of historical value are also dependent on environmental factors. Relatively intact gardens that are easy to research and document pose a straightforward resolution case for the designer. On the other hand, when territorial integrity has been impaired or the historic value transformed (built-in, overgrown, dilapidated, etc.) and fallen into ruins, the authentic renewal is a complex and responsible task since the monument conservation must properly be carried out alongside missing data supplementation or completion. Reconstruction of a garden or garden element can only occur if no other memory needs to be destroyed for its sake, and there are sufficient genuine pieces of information available. The Florence Charter mentions this: 'In principle, no one period should be given precedence over any other, except in exceptional cases where the degree of damage or destruction affecting certain parts of a garden may be such that it is decided to reconstruct it on the basis of the traces that survive or of unimpeachable documentary evidence" [20].

Landscape architecture is an applied science. According to the guidelines on the preparation of scientific publications in garden history, the results of the research are based on general preliminary studies of garden and landscape history, the research results and experiences of several decades, the exploration and analysis of authentic historical sources, and the site surveys and assessments.

*3.1. Thesis No 1: The Castle, the Castle Garden, and the Surrounding Landscape Altogether Represent a Single Artistic and Compositional Unit*

From the 100 investigated locations in 79 cases, during the field survey, we discovered strong visual connections still existing between the castle, the garden, and the landscape.

Neither the castle or the castle garden should be interpreted independently. They are a single unit, and all the man-made and natural elements of the castle garden—often specific elements beyond the garden boundaries—make part of this unit. The castle and the garden altogether represent a composition that is an integral part of a complex system developed on artistic, cultural, historical, ecological, and economic bases. When the estate complexes were established, the views and visual axes made use of the landscape potential, relying on architectural and artistic tools to explore.

*3.2. Thesis No 2: Visual Links Applied to Castle Gardens at a Landscape Scale may Be Divided into Two Main Categories: Eye-Catchers and Prospects*

The views were produced by creating effects with the aid of visual tools, and by delivering symbolic meanings and messages. Regarding the planned visual links between the landscape and the garden, two types are possible to distinguish: the "eye-catchers" and the "prospects".

The eye-catchers are structures that draw attraction even from a greater distance and are applied as focal points of a visual axis or determine a specific visual link. The eye-catchers—as distinctive elements of the landscape—determine the structural layout of landscape gardens. In the periods of the 19th-century Sentimentalism and Romanticism, characteristic structures and buildings in the garden were used not only for functional but also for spiritual, political, and aesthetic reasons [21–23].

The definition of a prospect as a picturesque scene resulted from the large scale composition of designed landscapes comprising the skyline of the natural or built environment, the vegetation, and the water surfaces [9].

The prospect is the virtual extension of the garden boundaries, the inclusion of the surrounding landscape into the garden scenery. Very often, the same landscape composition allows for the definition of several representative visual axes and several connections [10]. This is especially true for larger garden landscapes—at this point, we can refer to an excellent foreign example, the most extended European garden landscape, the English garden of Dessau-Wörlitz, where Edith Kresta mentions more than 300 visual axes applied as parts of the composition [24].

Out of 100 investigated locations in 61 cases, altogether, we found 139 built features (eye-catchers) on site, which proves the garden and landscape compositional role of eye-catchers and can serve as a basis for the renewal of the visual communication between garden and landscape. (Figure 2).

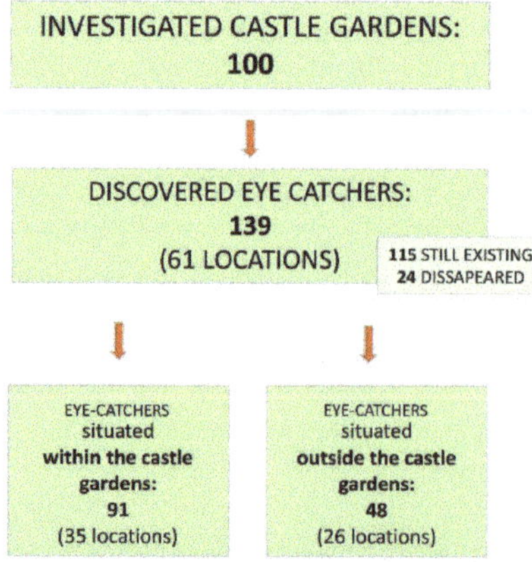


**Figure 2.** Eye-catchers location in Transylvanian castle gardens.

In several cases, we discovered former eye-catchers as well, which unfortunately disappeared in the course of the last century (Figure 3).

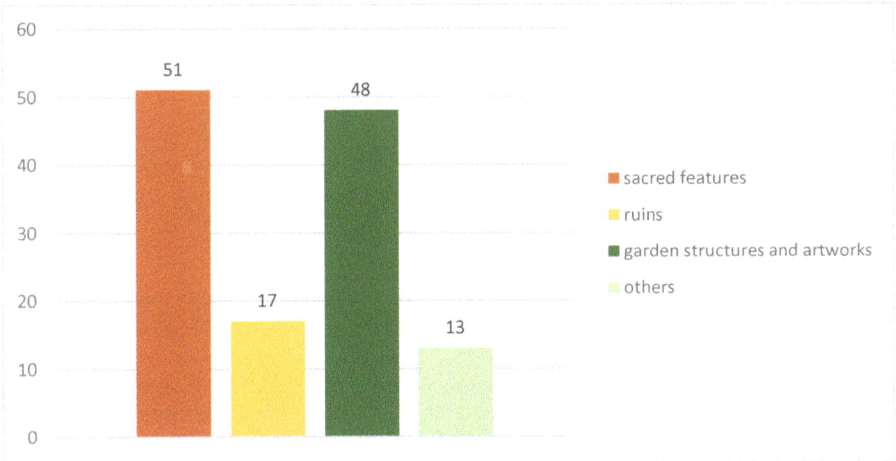

**Figure 3.** The Csicsal Hill in Magyarfenes (Vlaha) nowadays (photo by authors, 2018), without the Gloriette of the Jósika Castle Garden, a vanished eye-catcher represented by a sketch (bottom left) based on a photo (bottom right) [25].

*3.3. Thesis No 3: Eye-Catchers Are Predominantly Architectural Structures*

Eye-catchers applied as terminal points of planned visual axes of castle gardens are architectural or artistic works that draw attention even from a greater distance. They can be identified and classified (Figure 4).



**Figure 4.** Graphical representation of the main eye-catcher types defined during the Transylvanian Castel Garden Research and its numbers.

The eye-catchers applied most frequently in the castle gardens are:

- Sacred features/eye-catchers (church towers, burial monuments, tombs, memorial places, crypts, gloriettes, tempiettos, obelisks, etc.), altogether 51 (Figure 5);

- Ruins (of fortresses, castles, manor houses, churches, etc.), altogether 17;
- Garden structures and artworks (gazebos, pavilions, fountains, ornamental pools, cascades, grottos, garden cottages, flights of stairs, balustrades, sculptures, viaducts or other structures of staffage, and built garden features), altogether 48;
- Others, altogether 13.

(a)    (b)    (c)

**Figure 5.** Sacred eye-catchers: Bethlen Family Crypt, Nagyteremi (Tirimia, RO) (**a**); Memorial urn of József Teleki, Gernyeszeg (Gornesti, RO) (**b**); Pyramidal tomb of the Kemény Family, Csombord (Ciumbrud, RO) (**c**); Photos by Fekete, A, 2013 (**a**), 2016 (**b**), 2017 (**c**).

*3.4. Thesis No 4: Due to the Scarce Sources in Garden History, Garden Reconstruction Is Rarely a Feasible Option in Carpathian Basin. Other Types of Research-Based Design Approaches Have to Be Defined in Order to Ensure the Authentic and Value Preservation Oriented Historic Garden Renewals*

On the basis of the information available on the garden, the relevant heritage conservation principles, and the user needs, we have distinguished the following garden reconstruction categories for historic gardens:

- Garden renovation (revitalisation),
- Garden regeneration,
- Garden restoration.

## 4. Discussion

Results of the research carried out through several decades on historic gardens and landscapes can and ought to be applied to specific planning situations. That is why this chapter comprises selected examples[1] of projects where the historic value of the planned visual link between the garden and its surroundings is possible to identify, the compositional value of the visual axes and prospects is high and increases the aesthetic quality of the manor garden and the surrounding landscape; also, the visual features bear some symbolic meanings.

The projects for various sites introduce some specific historic garden reconstructions, which demonstrate the potential planning approaches required in a given situation for all garden reconstruction categories established. Through the examples, I intend to systematically summarise and evaluate the

---

[1] The authors plans realised.

garden reconstruction theories in Hungary at the beginning of the 21st century and to formulate new findings and proposals on the contemporary principles on historic garden restoration and planning.

The introductions of the renewal projects are not comprehensive: partly due to the limitations on length, and partly to focus on the topics (the garden scenery and the planned views) of the paper, only the relevant details of the projects have been included. Thus functional, technical and infrastructural, (roads, pavements, public utilities, lighting, earthworks), ecological and dendrological (plant geography, dendrology, plant associations, plant species applied, etc.), pedological, hydrological, hard landscaping, above ground drainage, and design aspects are introduced only to an extent necessary to have a general understanding of the specific example.

### 4.1. Garden Renovation (Revitalisation)

Garden reconstruction is based on the historical sources available, the heritage features of the site that are possible to identify, and the stylistic elements and analogies of a specific period. The park is renovated with a distinct application of the stylistic elements of the period most relevant to the site and with additional functions to meet the actual demands.

A good example of the renovation of the internal and external axial views is the Mikes Castle Garden in Zabola (Zabala, RO) [26].

The Mikes Castle in Zabola was built by Zsigmond Mikes in the 1620s as a two-storey Renaissance building with a quadratic layout. In the course of time, the castle underwent several renovations. Its current appearance dates back to 1867 when Benedek Mikes renovated the old building in Neoclassical style and extended it into three storeys. Both the castle and the 36 ha dendrological park that surrounds the building that were established in the 1880s in late landscape garden style are listed heritage assets. The property was nationalised in 1949, and both the buildings and the park significantly decayed in the second half of the 20th century. Elements of the listed heritage were irrecoverably damaged. The 2002 restitution returned the residence to the rightful owner, opening the opportunity for the renovation of the castle and the park.

In the course of the garden history research, it was possible to discover sources that could provide a basis for the reconstruction of the original spatial structure that had been only partially preserved by the beginning of the 21st century.

Two plans served as that foundation for the renovation of visual links. The construction of the park was started by the owner, Count Benedek Mikes in the 19th century, thereafter continued and completed by his son Ármin Mikes. A garden masterplan scaled 1:1000 and named *Blatt zur Konzeption II.* has been preserved from this period (Figure 6a). Although some of the details have never been realised, other details justify the appropriateness of the design concept, even nowadays. The title of the plan also hints at there having been a first development concept (presumably named *Blatt zur Konzepzion I.)*, which has not been found. A key person of the creation of the castle garden was Achille Duchene[2], a renowned French landscape architect [27].

Matched against the historic photographs, the old plans revealed the full layout of the terraces, which provided ground for the then functional (tennis courts, leisure grounds) and ornamental (rose garden, carpet beds) units, and also organised the visual links (Figures 6b and 7).

---

[2]  Achille Duchene (1866–1947), was a master of French geometric gardens in the first decades of the 20th century. His works may be found worldwide. He designed formal landscape gardens with distant and framed views. Main sites of his projects include: The Wigwam Garden, Villa Ephrussi de Rotschild, The Carolands, Courances Garden, Vaux le Vicomte Garden, Champs sur Marne, Courances and others. Also, he was the one who made the restoration plan for the water parterres of Blenheim Palace in England.

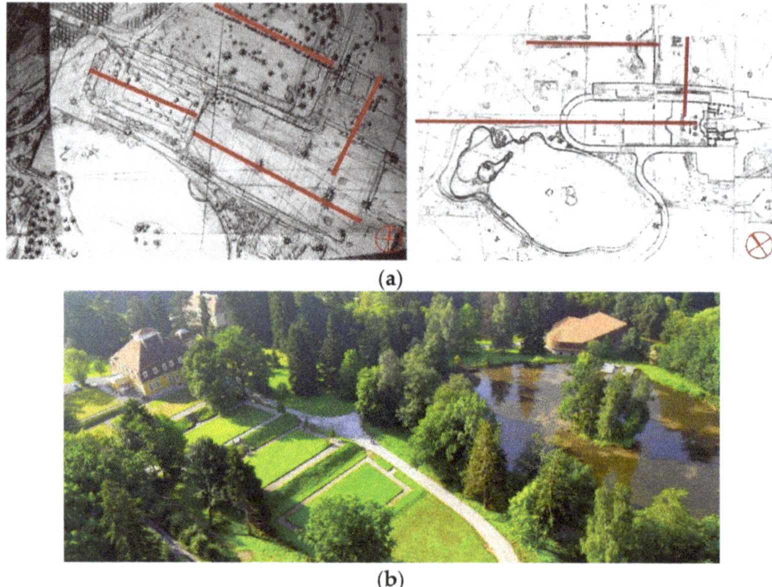

(a)

(b)

**Figure 6.** Renovation of terraces (**b**) and visual connections in the case of the Mikes Castle Garden in Zabola (Zabala, RO), based on analysis of old plans (**a**) and field survey. Design: Fekete, A.; Magdó, J.; Jávori, K. 2008–2010. Photo by the Mikes Family, 2014.

(a)                                    (b)

**Figure 7.** Terrace renewal in a minimalistic style, Mikes Castle Garden, Zabola (Zabala, R.O.). Source: Photo Mikes Family Archive, 1911 (**a**); photo by Fekete, A., 2013 (**b**).

With reference to the original plans preserved, an authentic renovation plan was prepared, with emphasis on the reconstruction of internal and external visual links. The geometric system of the terraces has been completely reconstructed, with the original mass and visual links restored. Nevertheless, taking into consideration new user demands, the terraces have been renovated without restoring also the former ornamental elements (rose garden, carpet beds, trimmed evergreen, and deciduous hedges) of the garden, with only lawn-covered horizontal surfaces and slopes, using and renovating the old structures of stairs (monolith stairs and columns, balustrades) through a minimalistic garden regeneration. The heritage assessment has proved that the site fully meets the criteria for historic gardens; therefore, as necessary in such circumstances, principles of heritage conservation must be applied. The decision for garden regeneration as the most appropriate solution is supported by the following facts:

- Since the 2002 restitution of the manor, the original residential function is only partial (old castle);
- For long term economic sustainability, the estate must maintain an important new function (as tourist accommodation): several renovated buildings of the estate (former boiler room, stable, shed, powerhouse, Swiss house) and the new castle all provide touristic services;
- The terms of renovation grants have necessitated the provision of additional new functions (venue, cultural centre);
- Uses arising from the new functions result in increased seasonal and occasional (events) loads that the garden should be capable of coping with;
- Residential functions and hospitality services should also be spatially separated in the castle garden, which requires an appropriate structural and functional layout of the garden.

In addition to serving the new functions, the main objectives of the design programme established in cooperation with the client were to restore the authentic spatial layout, atmosphere, and visual links that were specific to late landscape gardens, and also to renovate the related water infrastructure and the geometric system of terraces from the first years of the 20th century, adjacent to the old castle.

In the first half of the 20th century, the castle garden included a complex of wetland habitats (several ponds linked by watercourses and various structures used for water management). As a result of inappropriate management, most of the water surface had disappeared by the second half of the century, and a shrubby, woody vegetation took over the area of the former ponds. The renovation of the pond system and the wetland habitats has not only reinstated the ecological equilibrium and biodiversity but has also regenerated the internal visual links of the manor garden across the waters. Multiple versions of plans were prepared for the reconstruction of the water infrastructure (Figure 8).

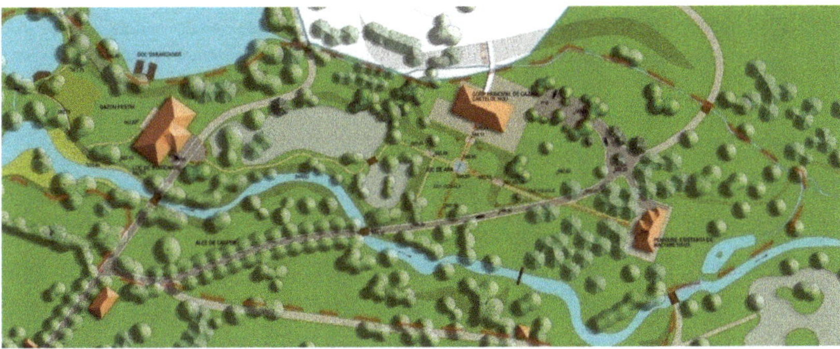

**Figure 8.** The final proposal for the revitalisation of the water infrastructure (detail), with the phases scheduled, Mikes Castle Garden, Zabola (Zabala, R.O.); Phase 1—blue; Phase 2—grey; design: Fekete, A.; Magdó, J.; Jávori, K. 2008–2010.

The overgrown shrubby vegetation and the water system has been cleared throughout the garden to a necessary extent, leading to the renovation of the spatial composition (Figure 9). The careful management of the woody plants has opened the space occupied by invasive plants for the dendrologically valuable tree specimens and also started the restoration of the generous views characteristic for landscape gardens. Long bygone visual axes and views important for spatial layout and composition have been revived, the connection between the garden and the landscape is seamless, boundaries dissolve, and views are undisturbed.

**Figure 9.** Renewed pondside views and visual axes through the water mirror in the Zabola Castle Garden. Photos by Fekete, A., 2018.

*4.2. Garden Regeneration*

The park/garden that has lost its historic features to a great extent is reconstructed on the basis of historical sources and analogies available. The solutions were applied to integrate the features of the relevant historical period into contemporary contexts. This approach establishes the opportunity of creating a contemporary work of art.

A good example of the historic garden regeneration is the case of the Regeneration of the Herb Garden and Arboretum of the Pannonhalma Archabbey, Hungary [28].

The project aimed at the regeneration of the 10-hectare Lavender Garden, and the related Herb Garden to the east of the Arboretum was also established around this time.

Similarly to the historical traditions of other orders, the Herb Garden always had an important role in Pannonhalma. Therefore, the introduction of the long traditions of the Pannonhalma Benedictines in medication and herb cultivation was a priority of the development, which has been complemented with the production of lavender oil in the past decades.

Our landscape design around the horticultural buildings of the archabbey recalls the atmosphere of traditional farmyards. The courtyard was fenced around by low stone walls, with a distinct separation of the outer versus inner areas, the farm buildings from the cultivated land. The show garden of herbs also reflects this approach. The strict geometry of the layout, the allocation of the planting beds, and the static water surface of the garden pool are all applied to strengthen the feeling of functionalism, while its scales recall the atmosphere of the historic monastery gardens. Nonetheless, with respect to the use of materials and the design of garden structures (pergola, ornamental pool), we applied contemporary solutions that are also capable of coping with the tourism load. Authentic depictions of the herb garden regarding its extent, exact location, and former design, which could have served for a direct restoration, were not available. That is why the garden was regenerated by analogies.

The layout, structures, and furniture of the garden are all geometric: benches are block-like, while the hard surfacing applies the same minor cobble stones and gravel that are general throughout the abbey. Retaining walls are all made with the grey limestone cover.

In the course of the planning, the reinforcement and highlighting of the relevant internal and external eye-catchers (the abbey tower, the new building of the distillery) and the views (towards the Arboretum and the surrounding landscape) had a priority. Faithfully to historical traditions, with the application of specific landscape design tools and accents in the composition, certain sections of the garden (e.g., the Herb Garden) were related distinctly to the abbey, while others (lavender fields) to the surrounding landscape (Figure 10).

(a)         (b)

**Figure 10.** Garden regeneration in the case of the herb garden in Pannonhalma Archabbey (HU). The view of the Cathedral's tower in the background suggest the sacred character of the site (**a**). The sub-Figure 11b shows the masterplan of the herb garden, marking with red line the renewed view towards the Cathedral's tower. Design: Fekete, A.; Vajda, Sz.; Szilágyi, K. 2010–2011. Photo by Fekete, A., 2014.

**Figure 11.** The restored pond, with the manor house in the background. Design: Fekete, A.; Rudd, M., Sárospataki, M., Weiszer, Á.; 2015. Photo by Fekete, A., 2017.

A second example of garden regeneration is the Kálnoky Castle Garden reconstruction in Miklósvár (Miclosoara, R.O.) [29]. The castle was built in Renaissance style in the 17th century. Despite several later Classical renovations, the building has preserved its Late Renaissance character. Owing to its structure and decoration, it is a noticeable monument in the Háromszék Region, reflecting representative functions of manor houses of Secler noblemen in the 17th and 18th centuries. The building is listed, managed by the former owner Kálnoky family.

Based on an analysis from the garden historical assessment criteria introduced in the doctoral thesis, the 14 ha manor park is a garden of historical value, and, according to descriptions and registers from the end of the 17th (1698) and beginning of the 18th (1716) centuries, its northern part adjacent to the manor house was a Renaissance garden with typical garden structures of the age. We have found clear descriptions or references to the pond, the gazebo on the peninsula, the wooden bridge, and the parterre flower garden [30].

The renovation of the garden was based on the following facts and findings:

- After the renovation, the use of the manor house and garden will be different from the original residential function;
- In the future, the function of the manor will be representative: partly as a cultural centre (a museum of the Transylvanian villages) and partly as a hotel;
- Uses arising from the new functions result in increased seasonal and occasional (events) loads, which the garden should be capable of coping with; the garden, as the first renewed Late Renaissance manor garden in Transylvania, should authentically recall the atmosphere of the 17th century Late Renaissance and serve as a good example for similar subsequent renovations.

In addition to serving the new functions, a main objective of the design programme established in cooperation with the client was to restore the functional units, layout, and axial views that were specific to Late Renaissance Transylvanian gardens. In this context, the renewal of the pond and the related garden structures (gazebo, wooden bridge) and the parterre garden was also proposed.

Based on historic records discovered during the garden history research, original functional units, layout, and axial views dating back to the 17th century were possible to restore so that the new functions were also properly supplied.

For the restoration of the pond and the retaining walls, beyond the evidence of written records, landform analysis has also confirmed the location of the pond during the site survey. Fed by a creek, the pond was located at the lowest part of the site, close to the creek.

As an open surface with visual links, the pond is an important element of the spatial structure of the garden, providing emphasis to the built features and supporting planned views. Its reflective surface also heightens the visual effects and experience (Figure 11). At the same time, the pond also served for economic purposes and used as a fish pond as it was typical for ponds in manor gardens in this period.

The castle is located at an elevation of 2.5 m above the pond behind the house. This position is beneficial not only for flood prevention and static reasons but also for the appearance and garden views: providing an accent to the castle that is, therefore, visible also from beyond the garden.

The levels of the castle and the pond were connected with a double line of parallel retaining walls. In the course of time, similarly to the dried bottom of the pond, the retaining walls also became filled and covered. Their line was, however, possible to be traced by the landform. The original location and structure of the retaining walls and the pond bottom were identified through archaeological survey, using an archaeological trench. The reconstruction of the retaining walls was based on the excavation and study of the structures well preserved underground. (Figure 12).

**Figure 12.** The castle facade towards the pond with the retaining walls in 2014 (**a**) and after the reconstruction in 2016 (**b**). The archaeological excavation of the retaining wall (**c**) and the renewed retaining wall (**d**). Photos by Fekete, A, 2017.

The gazebo is a pavilion-like roofed garden structure, mentioned in the 16th and 17th-century garden descriptions from Northern Hungary, Southern Hungary, and Transylvania [31].

The gazebo as an eye-catcher was also a basic feature in the late renaissance garden. The regeneration design of it was based on descriptions of gazebos from other Late Renaissance gardens in both Transylvania and the historical Upper Hungary, and also on the study of the characteristic elements of the 17th-century Transylvanian architecture, the wooden belfries that have been preserved at various places.

Despite the fact that written sources of the octagonal gazebo are scarce, it was possible to reconstruct it so that the size, shape, structural details, and materials are authentic (Figure 13).

**Figure 13.** View and technical details of the gazebo in Miklósvár, constructed according to historical descriptions and analogical examples and used as an eye-catcher. Photo by Fekete, A., 2017.

*4.3. Garden Restoration*

Garden reconstruction was based on the exploration of historical sources, the archaeological surveys, the use of preserved garden features, etc. Based on the preserved historic features and the available historical sources, it is possible to restore a part or the whole of the garden so that it will be greatly identical to the original design.

A good example in this category is represented by the restoration of the Upper Garden of the Royal Palace in Gödöllő, Hungary, which is one of the most valuable gems of Baroque in Hungary [32]. Construction of the Palace at Gödöllő was launched in 1735, under the direction of András Mayerhoffer. A unique feature of the Palace is the "inside out" nature of the layout, as the cour d'honneur does not serve as a reception area but looks towards the garden. Embracing the representative courtyard, the U-shaped building of the Palace provides a concave shape from the direction of the garden, which is ideal for a Baroque spatial composition.

The restoration strategy—following on from the above, and also with regard to the current and future functions of the garden—aimed at the restoration of the spatial composition and atmosphere that served for representation and recreation in the 19th century. Although the landscape garden has been reconstructed several times during its nearly 200-year history, the spatial layout and composition of the picturesque garden from the beginning of the 19th century has been preserved to a significant extent. Just as in the picturesque garden some significant features from the earlier Baroque have been retained, such as the pavilion on the Royal Mound, the horse chestnut alley, and the Shooting Gallery. In the last third of the century, promoted to a royal residence, Gödöllő became an exemplar of cultural and artistic representation of the age throughout the country. In the Romantic landscape garden established during this time, the naturalistic spatial layout and composition has been kept, while the use of plants was also moderate compared to the often rampant diversity of collector's gardens.

Regarding the planning, it was, therefore, necessary to go beyond the usual restoration clichés and concepts with an overwhelming priority of functional aspects. An artistic renovation based on the historical plant material and composition principles was the appropriate solution (Figure 14).

(a)  (b)

**Figure 14.** The restoration of visual links is related to a successful reconstruction of the spatial layout of the garden. The spatial layout is mainly determined by the vegetation, paths, and landforms. The restored path network (**b**) entirely corresponds to the one that existed in the Romantic landscape garden (**a**) in the 2nd half of the 19th century (layouts from 1867). Design by Fekete, A.; Jámbor, I.; Sárospataki, M.; Szilágyi, K.; Vajda, Sz. 2009–2010.

The restoration of the Upper Garden in Gödöllő was based on the pictorial and written sources available. A grass surfaced pleasure ground with an approximately 300-meter long visual axis provides an appropriate vista for the impressive building of the palace. Part of the garden near the palace is, thus, visually open since this is essential for the perception of the spatial layout and scales. Regarding other parts of the garden, the plan proposed the restoration of the historic network of paths. The romantic spatial composition with groves and confined spaces here and wide open views there, which was one characterised by the most refined English style, has been created with the use of vegetated boundaries and ornamental plantings [33,34] (Figure 15b,c).

(a)  (c)

**Figure 15.** (**a**) The restoration of the 400-m long horse chestnut alley, defining the baroque vista from the beginning of 18th century. 16. **b**, **c**. Colourful perennial boundaries, acting as space-defining elements and as visual focus points, too. Photos by Fekete, A, 2012 (**a**), 2016 (**b,c**).

Regarding internal visual links, determining a transverse visual axis, the Baroque pavilion on the Royal Mound is of outstanding importance. The full body seated bronze statue of Maria Theresa was placed on this axis as an eye-catcher.

The completely renovated, almost 400-m long horse chestnut alley of 120 trees is also of Baroque origin and determines a direction and vista parallel to the main axis, closed by the statue of Hercules with a cudgel in his hands. The alley starts from a flight of stairs made of monolith granite porphyry, flanked by short walls (Figure 15a).

## 5. Conclusions

In Hungary and especially in Transylvania, the current conditions of historic gardens and the scarce availability of historical sources on garden history in most of the cases do not allow for a full restoration of the original design. Despite this, the restoration or renovation of our historic castle gardens is an actual necessity.

Taking into consideration the driving forces (of natural or human origin) behind the change of landscapes and gardens, the original conditions of planned visual links of a garden may change during the times. A bi-directional visual link may become unidirectional or completely perish. The main factors that can have an impact on visual links may be of natural or human origin. Most of all, the loss of the former eye-catcher, the absence (disappearance, damage, relocation) of the architectural/artistic work, is the reason behind the change. Sometimes the change or the loss of a visual link may be the result of natural processes (and the lack of human intervention), for example, with the development of a forest cover. Occasionally, the modified or lost visual link is possible to restore during the landscape and garden restorations. Eye-catchers not only provide aesthetic experience but, through their positions and functions, also deliver messages. They are objects of symbolic occupation of space and represent national identity.

If scarce authentic information is at hand, contemporary tools are applied in order to evoke the atmosphere of the former garden, or the possibility of creating a new work arises.

Renovation should increase the value of the garden, ensuring the authenticity of the restoration. What is decisive during the process of restoration in the case of historic gardens is the application of the original spatial structure, spatial- and mass proportions, the restoration of the contemporary uniform composition.

During the restoration of the built elements (roads, pavements, garden structures, fixtures, etc.), the correct use of materials and traditional technologies is important. The distinction between original and restored elements, however, is expected to be unaffected even when using the same material, based on the principle of monumental authenticity. During the restoration of historic gardens, the application of the original plant species may prove difficult due to newly emerging pathogens or to the often limited variety of plants readily available. In this case, the use of substitute plant species is permitted.

The picturesqueness and the visually encoded message of the castle gardens, as well as the visual relationship between the castle estate and the surrounding landscape, are important for the conservation and restoration of our historic cultural landscapes. Owing to their visual features, artistic compositions, agelong processes of existence, and messages delivered, historic gardens bear a significant emotional substance. Their restoration is important also for experiencing natural processes. Redefinition of the relationships between the society and the heritage, humankind, and the landscape is a duty of all periods, and it urges us to always find the appropriate tools to do so.

Despite the difficulties in exploring their past, from the surveys of the actual conditions, it is possible to establish that our historic castle gardens still represent a part of our garden heritage that is possible to restore in the most authentic way. Regarding garden reconstructions, if we have no sufficient authentic information available on the original conditions, then there is the opportunity to recall the atmosphere of the former garden or occasionally to create a contemporary work of art.

The value of the presented renewals lies in the vast experience gained through the preparation of the inventor, which comprises a garden and landscape history review of nearly 400 years and complex documentation of numerous sites. Results highlight that manor estates were complex and functional units regarding their role in arts, society, and economy. Through their establishment, the maintenance, and the various agricultural activities contributing to economic sustainability, they had significant impacts on landscapes. The planners and owners made them an important tool and element of the transformation of the 18th- and 19th-century landscapes, and they still play a significant role in the character of the landscape nowadays. Their survey and documentation has an inestimable value for heritage conservation and is absolutely necessary for any future restorations. The research project won the Europa Nostra award in the research category in 2014, and the assessment by the jury highlights that the subject is one of great importance and has revealed something of a gap in our understanding of European garden history. The outcome provides the means not only to tackle the problems of decline and dereliction in the castle gardens themselves but also to learn ways in which other neglected gardens and garden-landscape relations in other regions can be rescued and restored [35].

**Author Contributions:** Conceptualization, A.F.; methodology, A.F. and L.K.; software, L.K.; validation, A.F. and L.K.; formal analysis, A.F.; investigation, A.F.; resources, A.F. and L.K.; data curation, A.F.; writing—original draft preparation, A.F.; writing—review and editing, A.F. and L.K.; visualization, A.F. and L.K.; supervision, A.F. and L.K.; project administration, A.F.; funding acquisition, A.F., L.K.

**Acknowledgments:** This material was published using the generous support of Fábos Foundation. Any opinions, findings, and conclusions expressed in this material are those of the authors and do not necessarily reflect the views of the Foundation.

**Conflicts of Interest:** The authors declare no conflict of interest.

## References

1.    Fekete, A. Vistas, Views, Prospects. Research and Design. In Proceedings of the Fábos Conference on Landscape and Greenway Planning, Amherst, MA, USA, 30–31 March 2019; 2019; Volume 6. Available online: https://scholarworks.umass.edu/fabos/vol6/iss1/53 (accessed on 25 October 2019). [CrossRef]

2.    Van den Brink, A.; Bruns, D.; Tobi, H.; Bell, S. *Research in Landscape Architecture. Methods and Methodology*; Routledge: New York, NY, USA, 2017; p. 316.

3.    Fekete, A. *Komponált látványok (Composed Vistas)*; Szent István Egyetem, Tájépítészeti és Településtervezési Kar: Budapest, Hungary, 2018; p. 168.

4.    Herczeg, Á. *Szép és kies kertek. A Kora Reneszánsz Kertek Itáliában és Magyarországon (Early Renaissance Gardens in Italy and Hungary)*; Pro-Print: Székelyudvarhely, Romania, 2006; p. 256.

5.    Hirschfeld, C.L. *Theorie der Gartenkunst II*; Weidmanns, Erben und Reich: Leipzig, Germany, 1779; Available online: https://digi.ub.uni-heidelberg.de/diglit/hirschfeld1779/0001 (accessed on 20 October 2019).

6.    Mavis, B.; Lambert, D. *The English Garden Tour*; John Murray Ltd.: London, UK, 1990; p. 156.

7.    Fekete, A. *Transylvanian Garden Art. Castle Gardens along the Maros River*; Művelődés Műhely: Kolozsvár/Zilah, Romania, 2007; p. 175.

8.    Quest-Ritson, C. 2003: *The English Garden. A Social History*; Penguin Books: London, UK, 2001; p. 280.

9.    Buttlar von, A. *Der Landschaftsgarten Gartenkunst des Klassizismus und der Romantik*; DuMont Buchverlag GmbH. und Co.: Köln, Germany, 1989; p. 302.

10.   Fekete, A.; van den Toorn, M. The Maros River and its potential for landscape development. In *Greenways and Landscapes in Change, Proceedings of the 5th Fábos Conference on Landscape and Greenway Planning*; István, V., Sándor, J., Krisztina, F., Julius, G.F., Robert, L.R., Mark, S.L., László, K., Eds.; Szent István Egyetem Tájtervezési és Területfejlesztési Tanszék: Budapest, Hungary, 2016; pp. 333–341.

11.   Calnan, M.; Fretwell, K. *Rooted in History. Studies in Garden Conservation*; The National Trust Enterprises Ltd.: London, UK, 2001; p. 216.

12.   Szabó, T.A. Erdélyi Történeti Kertek egy Biológus Szemével (Historic Gardens in Transylvania described by a biologist). In *Galavics, Géza: Történeti Kertek (Historic Gardens)*; MTA Művészettörténeti Kutatóintézet-Mágus Kiadó: Budapest, Hungary, 2000; pp. 69–86,276.

13.   Fekete, A. *Az Erdélyi Kertművészet. Szamos Menti Kastélykertek (Transylvanian Garden Art. Castle Gardens along the Szamos River)*; Művelődés Műhely: Kolozsvár/Zilah, Romania, 2012; p. 120.

14. Fekete, A. *Kolozsvári Kertek. Kolozsvár Történeti Zöldfelületei (Gardens of Kolozsvár. Historic Green Areas of city of Kolozsvár)*; Művelődés Műhely: Kolozsvár/Zilah, Romania, 2004; p. 64.

15. Fekete, A. Designed Visual Connections in the Transylvanian Landscape. *Transylv. Nostra* **2013**, *7*, 39–48.

16. Fekete, A. Garden Culture and Approaches to Landscape from Gábor Bethlen to József II Teleki. In *Studies in the History of Early Modern Transylvania*; Romsics, I., Kiss, G.K., Eds.; Atlantic Studies on Society in Change No. 140; Columbia University Press: New York, NY, USA, 2011; pp. 396–419.

17. Everton, S. Why Is Rousham England's Most Influential Garden? Available online: https://www.gardensillustrated.com/gardens/why-is-rousham-englands-most-influential-garden/ (accessed on 29 October 2019).

18. Steenbergen, C.; Reh, W. *Architecture and Landscape. The Design Experiment of the Great European Gardens and Landscapes*; Prestel Verlag: Munich, Germany, 1996; p. 373.

19. Kóbori, D. A Gernyeszegi Kastélykert Megújítása. A Kastélykertek Szerepe a Maros Tere Táji Örökségében (Renewal of the Castle Garden from Gernyeszeg. The Role of the Castle Gardens in the Landscape Heritage of the Upper Maros Valley). Master's Thesis, Szent István University, Budapest, Hungary, 2018; p. 87.

20. The Florence Charter. Available online: https://www.icomos.org/charters/gardens_e.pdf (accessed on 29 October 2019).

21. Hajdu Nagy, G. Rusztikus Építmények a Magyar Kertművészetben—Romok, Grották, Remeteségek (Rustic Features in Hungarian Garden Art—Ruins, Grottos, Hermitages). Ph.D. Thesis, Budapesti Corvinus Egyetem, Tájépítészeti és Tájökológiai Doktori Iskola, Budapest, Hungary, 2011; p. 327.

22. Guralnik, D. *Webster's New World College Dictionary*, 4th ed.; Houghton Mifflin Harcourt: Boston, MA, USA, 2004; p. 1728.

23. Curl, J.S. *A Dictionary of Architecture and Landscape Architecture*, 2nd ed.; Oxford University Press: Oxford, UK, 2006; p. 912.

24. Kresta, E. *Nature by Design—Lush Gardens and Spacious Parks Characterize the Garden Kingdom of Dessau-Wörlitz*; The Atlantic Times, Times Media GmbH: Berlin, Germany, 2008.

25. Kelemen, L. *A Magyarfenesi Báró Jósika-Kastély (The Jósika Castle from Magyarfenes)*; Művészeti Szalon; Providentia Nyomda: Kolozsvár, Romania, 1927; Volume 12, pp. 3–10.

26. Fekete, A.; Magdó, J.; Jávori, K. *A Zabolai Mikes-Kastélypark Tájépítészeti Megújítási Terve (The Landscape Renewal Plan of the Mikes Castle Garden from Zabola)*, Landscape Architectural Documentation Prepared for the Mikes Family; Zabola (Zabala), Romania, 2008–2010. 120.

27. Duchene, A. *Les Jardins de L'Avenir*; Vincent, Fréal & Cie: Paris, France, 1935.

28. Fekete, A.; Vajda, S.Z.; Szilágyi, K. *A Pannonhalmi Bencés Főapátság Arborétumának és Gyógynövénykertjének megújítási terve (The Landscape Renewal Plan of the Arboretum and Herb Garden from the Archabbey of Pannonhalma, Hungary)*, Landscape Architectural Documentation Prepared for the Pannonhalma Archabbey; Budapest, Hungary. 2012; 68.

29. Fekete, A.; Rudd, M.; Sárospataki, M.; Weiszer, Á. *A Miklósvári Kálnoky Kastélykert Megújítási Terve (The Landscape Renewal Plan of the Kálnoky Castle Garden from Miklósvár)*, Landscape Architectural Documentation Prepared for the Kálnoky Foundation; Miklósvár (Miclosoara), Romania. 2015; 144.

30. Stirling, J. *Magyar Reneszánsz Kertművészet a XVI-XVII. Században (Hungarian Garden Art in the 16–17 Centuries)*; Enciklopédia Kiadó: Budapest, Hungary, 1996; p. 304.

31. Margit, B.N. *Várak, Kastélyok, Udvarházak, Ahogy a Régiek Látták (Ancient Fortresses, Castles and Manor Houses)*; Kriterion Kiadó: Bukarest, Romania, 1973; p. 406.

32. Fekete, A.; Jámbor, I.; Sárospataki, M.; Szilágyi, K.; Vajda, S.Z. *A Gödöllői Királyi Kastély Felső Kertjének Rehabilitációja (The Landscape Renewal Plan of the Upper Garden of the Royal Garden from Gödöllő)*, Landscape Architectural Documentation Prepared for the Gödöllő Királyi Kastély Kht; Budapest, Hungary, 2009–2010. 114.

33. Hobhouse, P. *Plants in Garden History*; Chrysalis: London, UK, 1992; p. 336.

34. Berger, E.; Marz, J. *Cultural Landscapes. A Working Bibliography*, 5th ed.; ICOMOS-IFLA International Scientific Committee on Cultural Landscapes (ISCCL): Nürnberg, Germany; Wien, Austria, 2010; p. 141.

35. Available online: http://www.europanostra.org/awards/139/ (accessed on 3 May 2016).

MDPI
St. Alban-Anlage 66
4052 Basel
Switzerland
Tel. +41 61 683 77 34
Fax +41 61 302 89 18
www.mdpi.com

*Land* Editorial Office
E-mail: land@mdpi.com
www.mdpi.com/journal/land